AMERICAN MATHEMATICAL SOCIE
COLLOQUIUM PUBLICATIONS
VOLUME XXXVI

TOPOLOGICAL DYNAMICS

BY

WALTER HELBIG GOTTSCHALK

ASSOCIATE PROFESSOR OF MATHEMATICS
UNIVERSITY OF PENNSYLVANIA

AND

GUSTAV ARNOLD HEDLUND

PROFESSOR OF MATHEMATICS
YALE UNIVERSITY

PUBLISHED BY THE
AMERICAN MATHEMATICAL SOCIETY
Providence, Rhode Island
1955

Reprinted with corrections and an appendix, 1968

Reprinted 1974 in the United States of America

ISBN 0-8218-1036-7

PREFACE

By *topological dynamics* we mean the study of transformation groups with respect to those topological properties whose prototype occurred in classical dynamics. Thus the word "topological" in the phrase "topological dynamics" has reference to mathematical content and the word "dynamics" in the phrase has primary reference to historical origin.

Topological dynamics owes its origin to the classic work of Henri Poincaré and G. D. Birkhoff. It was Poincaré who first formulated and solved problems of dynamics as problems in topology. Birkhoff contributed fundamental concepts to topological dynamics and was the first to undertake its systematic development.

In the classic sense, a dynamical system is a system of ordinary differential equations with at least sufficient conditions imposed to insure continuity and uniqueness of the solutions. As such, a dynamical system defines a (one-parameter or continuous) flow in a space. A large body of results for flows which are of interest for classical dynamics has been developed, since the time of Poincaré, without reference to the fact that the flows arise from differential equations. The extension of these results from flows to transformation groups has been the work of recent years. These extensions and the concomitant developments are set forth in this book.

Part One contains the general theory. Part Two contains notable examples of flows which have contributed to the general theory of topological dynamics and which in turn have been illuminated by the general theory of topological dynamics.

In addition to the present Colloquium volume, the only books which contain extensive related developments are G. D. Birkhoff [2, Chapter 7], Niemytzki and Stepanoff [1, Chapter 4 of the 1st edition, Chapter 5 of the 2nd edition] and G. T. Whyburn [1, Chapter 12]. The contents of this volume meet but do not significantly overlap a forthcoming book by Montgomery and Zippin.

The authors wish to express their appreciation to the American Mathematical Society for the opportunity to publish this work. They also extend thanks to Yale University and the Institute for Advanced Study for financial aid in the preparation of the manuscript. The second named author extends to the American Mathematical Society his thanks for the invitation to give the Colloquium Lectures in which some aspects of the subject were discussed. Some of his work has been supported by the United States Air Force through the Office of Scientific Research of the Air Research and Development Command.

PHILADELPHIA, PENNSYLVANIA
NEW HAVEN, CONNECTICUT
July, 1954

CONVENTIONS AND NOTATIONS

Each of the two parts of the book is divided into sections and each section into paragraphs. Cross references are to paragraphs. 4.6 is the sixth paragraph of section 4. In general, a paragraph is either a definition, lemma, theorem or remark. A "remark" is a statement, the proof of which is left to the reader. These proofs are not always trivial, however.

References to the literature are, in general, given in the last paragraph of each section. Numbers in brackets following an author's name refer to the bibliography at the end of the book. Where there is joint authorship, the number given refers to the article or book as listed under the first named author.

An elementary knowledge of set theory, topology, uniform spaces and topological groups is assumed. Such can be gained by reading the appropriate sections of Bourbaki [1, 2, 3]. With a few exceptions to be noted, the notations used are standard and a separate listing seemed unnecessary.

Unless the contrary is specifically indicated, groups are taken to be multiplicative. Topological groups are not assumed to be necessarily separated (Hausdorff). The additive group of integers will be denoted by $\mathcal{I}$ and the additive group of reals by $\mathcal{R}$.

Contrary to customary usage, the function or transformation sign is usually placed on the right. That is, if X and Y are sets, f denotes a transformation of X into Y and $x \in X$, then xf denotes the unique element of Y determined by x and f.

In connection with uniform spaces, the term *index* is used to denote an element of the filter defining the uniform structure, thus replacing the term *entourage* as used by Bourbaki [2]. In keeping with the notation for the value of a function, if X is a uniform space, α is an index of X and $x \in X$, then $x\alpha$ denotes the set of all $y \in X$ such that $(x, y) \in \alpha$. Unless the contrary is stated, a uniform space is not necessarily separated.

TABLE OF CONTENTS

PART I. THE THEORY

1. TRANSFORMATION GROUPS

1.01. DEFINITION. A *topological transformation group,* or more briefly, a *transformation group,* is defined to be an ordered triple (X, T, π) consisting of a nonvacuous topological space X, a topological group T and a mapping $\pi : X \times T \to X$ such that:

(1) (Identity axiom) $(x, e)\pi = x$ $(x \in X)$ where e is the identity element of T. This axiom may be replaced by the assumption that π is onto.

(2) (Homomorphism axiom) $((x, t)\pi, s)\pi = (x, ts)\pi$ $(x \in X; t, s \in T)$.

(3) (Continuity axiom) π is continuous.

If (X, T, π) is a transformation group, then $\{X\}\{T\}\{\pi\}$ is called the *phase* $\{$*space*$\}$ $\{$*group*$\}$ $\{$*projection*$\}$ *of* (X, T, π). The phase projection is also called the *action.*

1.02. DEFINITION. Let X, Y be $\{$topological$\}\{$uniform$\}$ spaces and let (X, T, π), (Y, S, ρ) be transformation groups.

A $\{$*topological*$\}\{$*uniform*$\}$ *isomorphism of* (X, T, π) *onto* (Y, S, ρ) is defined to be a couple (h, φ) consisting of a $\{$homeomorphism$\}\{$unimorphism$\}$ h of X onto Y and a homeomorphic group-isomorphism φ of T onto S such that $(xh, t\varphi)\rho = (x, t)\pi h$ $(x \in X, t \in T)$.

The transformation groups (X, T, π) and (Y, S, ρ) are said to be $\{$*topologically*$\}\{$*uniformly*$\}$ *isomorphic* (each *to* or *with* the other) provided there exists a $\{$topological$\}\{$uniform$\}$ isomorphism of (X, T, π) onto (Y, S, ρ).

1.03. DEFINITION. Let X be a $\{$topological$\}\{$uniform$\}$ space and let (X, T, π) be a transformation group. An *intrinsic* $\{$*topological*$\}\{$*uniform*$\}$ property of (X, T, π) is a property of (X, T, π) definable solely in terms of the $\{$topological$\}$ $\{$uniform$\}$ structure of X, the topological structure of T, the group structure of T, and the mapping π.

1.04. REMARK. We propose in this monograph to study certain intrinsic properties of transformation groups. It is clear that intrinsic $\{$topological$\}$ $\{$uniform$\}$ properties of transformation groups are invariant under $\{$topological$\}$ $\{$uniform$\}$ isomorphisms.

1.05. NOTATION. Let (X, T, π) be a transformation group. If $x \in X$ and if $t \in T$, then $(x, t)\pi$ is denoted more concisely by xt when there is no chance for ambiguity. Then the identity and homomorphism axioms may be restated as follows:

(1) $xe = x$ $(x \in X)$.

(2) $(xt)s = x(ts)$ $(x \in X; t, s \in T)$.

1.06. TERMINOLOGY. The statement "(X, T, π) is a transformation group" may be paraphrased as "T $\{$is$\}\{$acts as$\}$ a transformation group $\{$of$\}\{$on$\}$ X

with respect to π". By virtue of 1.05 it often happens that a symbol for the phase projection does not occur in a discussion of a transformation group. In such an event we may speak simply of (X, T) as the transformation group where X is the phase space, T is the phase group and the phase projection is understood. The statement "(X, T) is a transformation group" may be paraphrased as "T {is}{acts as} a transformation group {of}{on} X". Thus the transformation group (X, T, π) may be denoted by (X, T) or even by T provided no ambiguity can occur.

Generally speaking, the statement that the transformation group (X, T, π) has a certain property may be paraphrased as either T has the property *on* X or X has the property *under* T. If $x \in X$, then the statement that (X, T, π) has a certain property at x may be paraphrased as either T has the property *at* x or x has the property *under* T.

1.07. STANDING NOTATION. Throughout the remainder of this section (X, T, π) denotes a transformation group.

1.08. DEFINITION. If $t \in T$, then the *t-transition of* (X, T, π), denoted π^t, is the mapping $\pi^t : X \to X$ such that $x\pi^t = (x, t)\pi = xt$ $(x \in X)$. The *transition group of* (X, T, π) is the set $G = [\pi^t \mid t \in T]$. The *transition projection of* (X, T, π) is the mapping $\lambda : T \to G$ such that $t\lambda = \pi^t$ $(t \in T)$.

If $x \in X$, then the *x-motion of* (X, T, π), denoted π_x, is the mapping $\pi_x : T \to X$ such that $t\pi_x = (x, t)\pi = xt$ $(t \in T)$. The *motion space of* (X, T, π) is the set $M = [\pi_x \mid x \in X]$. The *motion projection of* (X, T, π) is the mapping $\mu : X \to M$ such that $x\mu = \pi_x$ $(x \in X)$.

1.09. DEFINITION. The transformation group (X, T) is said to be *effective* provided that if $t \in T$ with $t \neq e$, then $xt \neq x$ for some $x \in X$.

1.10. REMARK. Let {G}{λ}{M}{μ} be the {transition group}{transition projection}{motion space}{motion projection} of (X, T, π). Then

(1) π^e is the identity mapping of X.

(2) If $t, s \in T$, then $\pi^t\pi^s = \pi^{ts}$.

(3) If $t \in T$, then π^t is a one-to-one mapping of X onto X and $(\pi^t)^{-1} = \pi^{t^{-1}}$.

(4) If $t \in T$, then π^t is a homeomorphism of X onto X.

(5) G is a group of homeomorphisms of X onto X.

(6) λ is a group-homomorphism of T onto G. This justifies the name "homomorphism axiom" of 1.01 (2).

(7) λ is one-to-one if and only if (X, T, π) is effective.

(8) If $x \in X$, then π_x is a continuous mapping of T into X.

(9) μ is a one-to-one mapping of X onto M.

1.11. REMARK. Let $t \in T$ and let $\varphi_t : T \to T$ be defined by $\tau\varphi_t = t^{-1}\tau t$ $(\tau \in T)$. Then (π^t, φ_t) is a topological isomorphism of (X, T, π) onto (X, T, π).

1.12. DEFINITION. Let X be a topological space. A *topological homeomorphism group of* X is a topologized group Φ of homeomorphisms of X onto X

such that Φ is a topological group and $\rho : X \times \Phi \to X$ is continuous where ρ is defined by $(x, \varphi)\rho = x\varphi$ $(x \in X, \varphi \in \Phi)$.

1.13. Remark. The effective topological transformation groups and the topological homeomorphism groups are essentially identical in the following sense:

(1) If (X, T, π) is an effective topological transformation group, then the transition group G of (X, T, π), topologized so that the transition projection of (X, T, π) becomes a group-isomorphic homeomorphism of T onto G, is a topological homeomorphism group of X.

(2) If Φ is a topological homeomorphism group of a topological space X, then (X, Φ, ρ) is an effective topological transformation group where $\rho : X \times \Phi \to X$ is defined by $(x, \varphi)\rho = x\varphi$ $(x \in X, \varphi \in \Phi)$.

In particular, a notion defined for topological transformation groups is automatically defined for topological homeomorphism groups.

1.14. Definition. A *discrete transformation group* is a topological transformation group whose phase group is discrete. A *discrete homeomorphism group* is a topological homeomorphism group provided with its discrete topology. A *homeomorphism group* is a group of homeomorphisms. The *total homeomorphism group* of a topological space X is the group of all homeomorphisms of X onto X.

1.15. Remark. It is clear from 1.13 that the effective discrete transformation groups and the discrete homeomorphism groups are to be considered as identical. Since a homeomorphism group may be considered as a discrete homeomorphism group, a notion defined for transformation groups is automatically defined for homeomorphism groups.

1.16. Notation. If $A \subset X$ and if $B \subset T$, then $(A \times B)\pi = [xt \mid x \in A \& t \in B]$ is denoted more concisely by AB when there is no chance for ambiguity. In particular, we write At in place of $A[t]$ where $A \subset X$ and $t \in T$; and we write xB in place of $[x]B$ where $x \in X$ and $B \subset T$. By the homomorphism axiom, xts is unambiguously defined, where $x \in X$ and $t, s \in T$; likewise ABC where $A \subset X$ and $B, C \subset T$; etc.

1.17. Lemma. *Let X, Y, Z be topological spaces and let $\varphi : X \times Y \to Z$ be continuous. If A, B are compact subsets of X, Y and if W is a neighborhood of $(A \times B)\varphi$, then there exist neighborhoods U, V of A, B such that $(U \times V)\varphi \subset W$.*

Proof. We write $(x, y)\varphi = xy$ $(x \in X, y \in Y)$.

Let $x \in A$. We show there exist open neighborhoods U, V of x, B such that $UV \subset W$. For each $y \in B$ there exist open neighborhoods U_y, V_y of x, y such that $U_yV_y \subset W$. Choose a finite subset F of B for which $B \subset \bigcup_{y \in F} V_y$. Define $U = \bigcap_{y \in F} U_y$ and $V = \bigcup_{y \in F} V_y$.

For each $x \in A$ there exist open neighborhoods U_x, V_x of x, B such that $U_xV_x \subset W$. Choose a finite subset E of A for which $A \subset \bigcup_{x \in E} U_x$. Define $U = \bigcup_{x \in E} U_x$ and $V = \bigcap_{x \in E} V_x$. Then U, V are neighborhoods of A, B such that $UV \subset W$.

1.18. LEMMA. *The following statements are valid:*

(1) *If* $A \subset X$ *and if* $t \in T$, *then* $\overline{A}t = \overline{At}$.

(2) *If* $A \subset X$ *and if* $B \subset T$, *then* $\overline{A}\,\overline{B} \subset \overline{AB}$ *and* $\overline{\overline{A}B} = \overline{A\overline{B}} = \overline{AB}$.

(3) *If* A, B *are compact subsets of* X, T, *then* AB *is a compact subset of* X.

(4) *If* A, B *are compact subsets of* X, T *and if* W *is a neighborhood of* AB, *then there exist neighborhoods* U, V, *of* A, B *such that* $UV \subset W$.

(5) *If* A *is a closed subset of* X *and if* B *is a compact subset of* T, *then* AB *is a closed subset of* X.

(6) *If* $A \subset X$ *and if* B *is a compact subset of* T, *then* $\overline{AB} = \overline{A}B$.

PROOF. (1) Since $\pi^t : X \to X$ is a homeomorphism onto, $\overline{At} = \overline{A\pi^t} = \overline{A}\pi^t = \overline{A}t$.

(2) Since $\pi : X \times T \to X$ is continuous, $\overline{A}\,\overline{B} = (\overline{A} \times \overline{B})\pi = (\overline{A \times B})\pi \subset \overline{(A \times B)\pi} = \overline{AB}$. The last conclusion follows from $AB \subset \overline{A}B \subset \overline{AB}$ and $AB \subset A\overline{B} \subset \overline{AB}$.

(3) $AB = (A \times B)\pi$ is a continuous image of the compact set $A \times B$.

(4) Use 1.17.

(5) Let $x \in X - AB$. Then $xB^{-1} \cap A = \emptyset$. By (4) there exists a neighborhood U of x such that $\overline{UB^{-1}} \cap A = \emptyset$ whence $U \cap AB = \emptyset$ and $U \subset X - AB$.

(6) By (2) and (5), $\overline{AB} = \overline{\overline{A}B} = \overline{A}B$.

1.19. LEMMA. *Let* X, Y *be uniform spaces, let* $\varphi : X \to Y$ *be continuous and let* A *be a compact subset of* X. *If* β *is an index of* Y, *then there exists an index* α *of* X *such that* $x \in A$ *implies* $x\alpha\varphi \subset x\varphi\beta$.

PROOF. Let γ be an index of Y such that $\gamma^2 \subset \beta$. For each $x \in A$ there exists a symmetric index α_x of X such that $x\alpha_x^2\varphi \subset x\varphi\gamma$. Choose a finite subset E of A for which $A \subset \bigcup_{x\in E} x\alpha_x$. Define $\alpha = \bigcap_{x \in E} \alpha_x$. Let $x \in A$. There exists $z \in E$ such that $x \in z\alpha_z$. Since $x\varphi \in z\alpha_z^2\varphi \subset z\varphi\gamma$, it follows that $x\alpha\varphi \subset z\alpha_z\alpha\varphi \subset z\alpha_z^2\varphi \subset z\varphi\gamma \subset x\varphi\gamma^2 \subset x\varphi\beta$. The proof is completed.

1.20. LEMMA. *Let* X *be a uniform space, let* A, B *be compact subsets of* X, T *and let* α *be an index of* X. *Then:*

(1) *There exists an index* β *of* X *and a neighborhood* V *of* e *such that* $x \in A$ *and* $t \in B$ *implies* $x\beta tV \subset xt\alpha$ *and* $x\beta Vt \subset xt\alpha$.

(2) *There exists an index* β *of* X *such that* $x \in A$ *and* $t \in B$ *implies* $x\beta t \subset xt\alpha$ *and* $xt\beta \subset x\alpha t$.

(3) *There exists an index* β *of* X *such that* $x \in A$ *implies* $x\beta B \subset xB\alpha$ *and* $xB\beta \subset x\alpha B$.

(4) *There exists an index* β *of* X *such that* $t \in B$ *implies* $A\beta t \subset At\alpha$ *and* $At\beta \subset A\alpha t$.

PROOF. Since $\pi : X \times T \to X$ is continuous and $A \times B$ is a compact subset of $X \times T$, (1) follows from 1.19. The first part of (2) follows immediately from (1). Since AB and B^{-1} are compact, there exists an index β of X such that $x \in AB$ and $t \in B^{-1}$ implies $x\beta t \subset xt\alpha$. Hence, $x \in A$ and $t \in B$ implies $xt\beta t^{-1} \subset xtt^{-1}\alpha$

and $xt\beta \subset x\alpha t$. This proves the second part of (2). Finally (3) and (4) are easy consequences of (2).

1.21. LEMMA. *Let X be a compact uniform space, let α be an index of X and let K be a compact subset of T. Then there exists an index β of X such that:*

(1) *$x \in X$ and $k \in K$ implies $x\beta k \subset xk\alpha$.*

(2) *$x \in X$ and $k \in K$ implies $xk\beta \subset x\alpha k$.*

(3) *$(x, y) \in \beta$ and $k \in K$ implies $(xk, yk) \in \alpha$.*

(4) *$(x, y) \in \alpha'$ and $k \in K$ implies $(xk, yk) \in \beta'$.*

PROOF. Use 1.20 (2).

1.22. DEFINITION. Let $A \subset X$ and let $S \subset T$. The set A is said to be *invariant under* S or *S-invariant* provided that $AS \subset A$. When $S = T$, the qualifying phrase "under T" and the prefix "T-" may be omitted.

1.23. REMARK. The following statements are valid:

(1) If $A \subset X$, then the following statements are pairwise equivalent: A is T-invariant, that is, $AT \subset A$; $AT = A$; $t \in T$ implies $At \subset A$; $t \in T$ implies $At = A$; $t \in T$ implies $At \supset A$.

(2) X and $\emptyset$ are T-invariant.

(3) If A is a T-invariant subset of X, then $A' = X - A$, $\overline{A}$, int A are T-invariant.

(4) If A and B are T-invariant subsets of X, then $A - B$ is T-invariant.

(5) If $\mathcal{A}$ is a class of T-invariant subsets of X, then $\bigcap \mathcal{A}$ and $\bigcup \mathcal{A}$ are T-invariant.

(6) If $A \subset X$ and if $S \subset T$, then A is S-invariant if and only if A' is S^{-1}-invariant.

1.24. REMARK. Let $Y \subset X$, let S be a subgroup of T, let Y be S-invariant and let $\rho = \pi \mid Y \times S$. Then (Y, S, ρ) is a transformation group. In particular, T acts as a transformation group on every T-invariant subset of X, and every subgroup of T acts as a transformation group on X.

1.25. DEFINITION. Let Y be a T-invariant subset of X. The transformation group (X, T) is said to have a certain property *on* Y provided that the transformation group (Y, T) has this property.

1.26. DEFINITION. Let $x \in X$ and let $S \subset T$. The *orbit of x under S* or the *S-orbit of x* is defined to be the subset xS of X. The *orbit-closure of x under S* or the *S-orbit-closure of x* is defined to be the subset $\overline{xS}$ of X. An {*orbit*} {*orbit-closure*} *under* S or an {*S-orbit*} {*S-orbit-closure*} is defined to be a subset A of X such that A is the {S-orbit} {S-orbit-closure} of some point of X. When $S = T$, the phrase "under T" and the prefix "T-" may be omitted.

1.27. DEFINITION. Let X be a set. A *partition of* X is defined to be a disjoint class $\mathcal{A}$ of nonvacuous subsets of X such that $X = \bigcup \mathcal{A}$.

1.28. REMARK. The following statements are valid:

(1) If $x \in X$, then the orbit of x under T is the least T-invariant subset of X which contains the point x.

(2) If $x \in X$ and if $y \in xT$, then $yT = xT$.

(3) The class of all orbits under T is a partition of X.

(4) If $x \in X$, then the orbit-closure of x under T is the least closed T-invariant subset of X which contains the point x.

(5) If $x \in X$ and if $y \in \overline{xT}$, then $\overline{yT} \subset \overline{xT}$.

(6) The class of all orbit-closures under T is a covering of X.

1.29. REMARK. The following definitions describe various methods of constructing transformation groups.

1.30. DEFINITION. Let n be a positive integer. An *n-parameter* {*discrete*} {*continuous*} *flow* is defined to be a transformation group whose phase group is $\{\mathcal{I}^n\}\{\mathcal{R}^n\}$. The phrase "one-parameter {discrete}{continuous} flow" is shortened to "{discrete}{continuous} flow".

1.31. REMARK. Let n be a positive integer. An n-parameter discrete flow $(X, \mathcal{I}^n, \pi)$ is characterized in an obvious manner by n pairwise commuting homeomorphisms of X onto X, namely $\pi^{(1,0,\cdots,0)}, \cdots, \pi^{(0,\cdots,0,1)}$ which are said to *generate* $(X, \mathcal{I}^n, \pi)$. In particular, a discrete flow $(X, \mathcal{I}, \pi)$ is characterized by a single homeomorphism of X onto X, namely π^1, which is said to *generate* $(X, \mathcal{I}, \pi)$. The properties of a discrete flow $(X, \mathcal{I}, \pi)$ are often attributed to its generating homeomorphism π^1.

1.32. DEFINITION. Let S be a subgroup of T and define $\rho = \pi \mid X \times S$. The transformation group (X, S, ρ) is called the *S-restriction of* (X, T, π) or a *subgroup-restriction of* (X, T, π).

Let Y be a subset of X such that $(Y \times T)\pi = Y$ and define $\rho = \pi \mid Y \times T$. The transformation group (Y, T, ρ) is called the *Y-restriction of* (X, T, π) or a *subspace-restriction of* (X, T, π).

Let S be a subgroup of T, let Y be a subset of X such that $(Y \times S)\pi = Y$ and define $\rho = \pi \mid Y \times S$. The transformation group (Y, S, ρ) is called the *(Y, S)-restriction of* (X, T, π) or a *transformation subgroup of* (X, T, π).

Let S be a topological group, let $\varphi : S \to T$ be a continuous homomorphism into and let $\rho : X \times S \to X$ be defined by $(x, s)\rho = (x, s\varphi)\pi$ $(x \in X, s \in S)$. The transformation group (X, S, ρ) is called the *(S, φ)-restriction of* (X, T, π).

Let S be a topological group, let $\varphi : S \to T$ be a continuous homomorphism into, let Y be a subset of X such that $(Y \times S\varphi)\pi = Y$ and let $\rho : Y \times S \to Y$ be defined by $(y, s)\rho = (y, s\varphi)\pi$ $(y \in Y, s \in S)$. The transformation group (Y, S, ρ) is called the *(Y, S, φ)-restriction of* (X, T, π).

1.33. REMARK. We consider every partition $\mathcal{A}$ of a topological space X to be itself a topological space provided with its partition topology, namely, the greatest topology which makes the projection of X onto $\mathcal{A}$ continuous.

1.34. Definition. Let X be a set, let $\mathcal{A}$ be a partition of X and let $E \subset X$. The *star of E in $\mathcal{A}$* or the *$\mathcal{A}$-star of E* or the *saturation of E in $\mathcal{A}$* or the *$\mathcal{A}$-saturation of E*, denoted $E\mathcal{A}$, is the subset $\bigcup[A \mid A \in \mathcal{A},\ A \cap E \neq \emptyset]$ of X. The set E is *saturated in $\mathcal{A}$* or *$\mathcal{A}$-saturated* in case $E = E\mathcal{A}$.

1.35. Definition. Let X be a topological space and let $\mathcal{A}$ be a partition of X. The partition $\mathcal{A}$ is said to be {*star-open*}{*star-closed*} provided that the $\mathcal{A}$-star of every {open}{closed} subset of X is {open}{closed} in X.

1.36. Remark. Let X be a topological space and let $\mathcal{A}$ be a partition of X. Then the following statements are pairwise equivalent:

(1) $\mathcal{A}$ is {star-open}{star-closed}.

(2) If $x \in X$ and if U is a neighborhood of $\{x\}\{x\mathcal{A}\}$, then there exists a neighborhood V of $\{x\mathcal{A}\}\{x$ and therefore $x\mathcal{A}\}$ such that $\{V \subset U\mathcal{A}\}\{V\mathcal{A} \subset U\}$.

(3) The projection of X onto $\mathcal{A}$ is {open}{closed}.

1.37. Definition. Let X be a topological space. A *decomposition of X* is a partition $\mathcal{A}$ of X such that every member of $\mathcal{A}$ is compact.

1.38. Remark. Let X be a compact metrizable space and let $\mathcal{A}$ be a decomposition of X. Then $\mathcal{A}$ is {star-open}{star-closed} if and only if $x_0, x_1, x_2, \cdots \in X$ with $\lim_{n\to\infty} x_n = x_0$ implies

$$\{x_0\mathcal{A} \subset \liminf_{n\to\infty} x_n\mathcal{A}\}\{\limsup_{n\to\infty} x_n\mathcal{A} \subset x_0\mathcal{A}\}.$$

1.39. Definition. Let $\mathcal{A}$ be a {star-open partition}{star-closed decomposition} of X, let $A\pi^t \in \mathcal{A}$ $(A \in \mathcal{A}, t \in T)$ and let $\rho : \mathcal{A} \times T \to \mathcal{A}$ be defined by $(A, t)\rho = A\pi^t$ $(A \in \mathcal{A}, t \in T)$. The transformation group $(\mathcal{A}, T, \rho)$ is called the *partition transformation group of $\mathcal{A}$ induced by* (X, T, π).

1.40. Definition. Let Φ be a group of homeomorphisms of X onto X such that $\varphi\pi^t = \pi^t\varphi$ $(\varphi \in \Phi, t \in T)$ and let $\mathcal{A} = [x\Phi \mid x \in X]$ whence $\mathcal{A}$ is a star-open partition of X such that $A\pi^t \in \mathcal{A}$ $(A \in \mathcal{A}, t \in T)$. The partition transformation group of $\mathcal{A}$ induced by (X, T, π) is called the *Φ-orbit partition transformation group induced by* (X, T, π).

1.41. Definition. Let Φ be a group of homeomorphisms of X onto X such that $\varphi\pi^t = \pi^t\varphi$ $(\varphi \in \Phi, t \in T)$ and let $\mathcal{A} = [\overline{x\Phi} \mid x \in X]$ be a partition of X whence $\mathcal{A}$ is a star-open partition of X such that $A\pi^t \in \mathcal{A}$ $(A \in \mathcal{A}, t \in T)$. The partition transformation group of $\mathcal{A}$ induced by (X, T, π) is called the *Φ-orbit-closure partition transformation group induced by* (X, T, π).

1.42. Definition. {Let φ be a continuous-open mapping of X onto X}{Let X be a compact T_2-space, let φ be a continuous mapping of X onto X} such that $\varphi\pi^t = \pi^t\varphi$ $(t \in T)$ and let $\mathcal{A} = [x\varphi^{-1} \mid x \in X]$ whence $\mathcal{A}$ is a {star-open partition}{star-closed decomposition} of X such that $A\pi^t \in \mathcal{A}$ $(A \in \mathcal{A},$

$t \in T$). The partition transformation group of $\mathfrak{A}$ induced by (X, T, π) is called the *φ-inverse partition transformation group induced by* (X, T, π).

1.43. DEFINITION. Let S be a topological group, let T be a topological subgroup of S, for $x \in X$ and $\sigma \in S$ define $\{A(x, \sigma) = [(x\pi^\tau, \sigma\tau) \mid \tau \in T]\}$ $\{A(x, \sigma) = [(x\pi^\tau, \tau^{-1}\sigma) \mid \tau \in T]\}$, define the star-open partition $\mathfrak{A} = [A(x, \sigma) \mid x \in X, \sigma \in S]$ of $X \times S$ and let $\rho : \mathfrak{A} \times S \to \mathfrak{A}$ be defined by $\{(A(x, \sigma), s)\rho = A(x, s^{-1}\sigma)\}\{(A(x, \sigma), s)\rho = A(x, \sigma s)\}$ $(x \in X; \sigma, s \in S)$. The transformation group $(\mathfrak{A}, S, \rho)$ is called the *{left}{right} S-extension of* (X, T, π).

1.44. REMARK. We adopt the notation of 1.43. Consider the transformation group $(X \times S, S, \eta)$ where $\eta : (X \times S) \times S \to X \times S$ is defined by $\{((x, \sigma), s)\eta = (x, s^{-1}\sigma)\}\{((x, \sigma), s)\eta = (x, \sigma s)\}$ $(x \in X; \sigma, s \in S)$. The partition transformation group of $\mathfrak{A}$ induced by $(X \times S, S, \eta)$ coincides with the {left}{right} S-extension of (X, T, π).

1.45. REMARK. Let S be a topological group and let T be a discrete topological subgroup of S. Then (X, T, π) is isomorphic to a transformation subgroup of the {left}{right} S-extension of (X, T, π).

1.46. NOTATION. The cartesian product of a family $(X_\iota \mid \iota \in I)$ of sets is denoted $\mathsf{X}_{\iota \in I} X_\iota$. The direct product of a family $(G_\iota \mid \iota \in I)$ of groups is denoted ${}_D\mathsf{X}_{\iota \in I} G_\iota$.

1.47. REMARK. We consider the cartesian product of every family of {topological}{uniform} spaces to be itself a {topological}{uniform} space provided with its product {topology}{uniformity}, namely, the least {topology}{uniformity} which makes all the projections onto the factor spaces {continuous} {uniformly continuous}.

1.48. DEFINITION. Let $((X_\iota, T_\iota, \pi_\iota) \mid \iota \in I)$ be a family of transformation groups. The *{cartesian}{direct} product of* $((X_\iota, T_\iota, \pi_\iota) \mid \iota \in I)$, denoted $\{\mathsf{X}_{\iota \in I} (X_\iota, T_\iota, \pi_\iota)\}$ $\{{}_D\mathsf{X}_{\iota \in I} (X_\iota, T_\iota, \pi_\iota)\}$, is the transformation group (X, T, π) where $X = \mathsf{X}_{\iota \in I} X_\iota$, $\{T = \mathsf{X}_{\iota \in I} T_\iota\}\{T = {}_D\mathsf{X}_{\iota \in I} T_\iota\}$ and $\pi : X \times T \to X$ is defined by $(x, t)\pi = (x_\iota \pi_\iota^{t_\iota} \mid \iota \in I)$ $(x = (x_\iota \mid \iota \in I) \in X, t = (t_\iota \mid \iota \in I) \in T)$.

1.49. DEFINITION. Let $((X_\iota, T, \pi_\iota) \mid \iota \in I)$ be a family of transformation groups. The *space product of* $((X_\iota, T, \pi_\iota) \mid \iota \in I)$, denoted ${}_S\mathsf{X}_{\iota \in I} (X_\iota, T, \pi_\iota)$, is the transformation group (X, T, π) where $X = \mathsf{X}_{\iota \in I} X_\iota$ and $\pi : X \times T \to X$ is defined by $(x, t)\pi = (x_\iota \pi_\iota^t \mid \iota \in I)$ $(x = (x_\iota \mid \iota \in I) \in X, t \in T)$.

1.50. REMARK. Both the direct and space products of a family of transformation groups are subgroup-restrictions of the cartesian product of the family.

1.51. DEFINITION. Let T be a topological group.

The *left transformation group of* T is defined to be the transformation group (T, T, λ) where $\lambda : T \times T \to T$ is defined by $(\tau, t)\lambda = t^{-1}\tau (\tau, t \in T)$.

The *right transformation group of* T is defined to be the transformation group (T, T, μ) where $\mu : T \times T \to T$ is defined by $(\tau, t)\mu = \tau t$ $(\tau, t \in T)$.

The *bilateral transformation group of* T is defined to be the transformation group $(T, T \times T, \xi)$ where $\xi : T \times (T \times T) \to T$ is defined by $(\tau, (t, s))\xi = t^{-1}\tau s$ $(\tau, t, s \in T)$.

1.52. DEFINITION. Let X be a uniform space and let $x \in X$. The transformation group (X, T, π) is said to be {*equicontinuous at* x}{*equicontinuous*}{*uniformly equicontinuous*} provided that the transition group $[\pi^t \mid t \in T]$ is {equicontinuous at x}{equicontinuous}{uniformly equicontinuous}.

1.53. REMARK. Let X be a uniform space. The following statements are pairwise equivalent:

(1) (X, T, π) is uniformly equicontinuous.

(2–5) If α is an index of X, then there exists an index β of X such that

$$\{x \in X \text{ and } t \in T \text{ implies } x\beta t \subset xt\alpha\}$$
$$\{x \in X \text{ and } t \in T \text{ implies } xt\beta \subset x\alpha t\}$$
$$\{(x, y) \in \beta \text{ and } t \in T \text{ implies } (xt, yt) \in \alpha\}$$
$$\{(x, y) \in \alpha' \text{ and } t \in T \text{ implies } (xt, yt) \in \beta'\}.$$

1.54. REMARK. The {left}{right} transformation group of a topological group is uniformly equicontinuous relative to the {left}{right} uniformity of the phase space.

1.55. NOTATION. Let H be a subgroup of a group G. The left quotient space $[xH \mid x \in G]$ of G by H is denoted G/H. The right quotient space $[Hx \mid x \in G]$ of G by H is denoted $G\backslash H$.

1.56. DEFINITION. Let S be a subgroup of a topological group T.

The *left transformation group of* T/S *induced by* T is defined to be the transformation group $(T/S, T, \lambda)$ where $\lambda : T/S \times T \to T/S$ is defined by $(A, t)\lambda = t^{-1}A$ $(A \in T/S, t \in T)$.

The *right transformation group of* $T\backslash S$ *induced by* T is defined to be the transformation group $(T\backslash S, T, \mu)$ where $\mu : T\backslash S \times T \to T\backslash S$ is defined by $(A, t)\mu = At$ $(A \in T\backslash S, t \in T)$.

1.57. REMARK. Let S be a subgroup of a topological group T and let (T, T, η) be the {left}{right} transformation group of T. Then the {left}{right} transformation group of $\{T/S\}\{T\backslash S\}$ induced by T coincides with the partition transformation group of $\{T/S\}\{T\backslash S\}$ induced by (T, T, η).

1.58. DEFINITION. Let φ be a continuous homomorphism of a topological group T into a topological group S and let $\rho : S \times T \to S$ be defined by $\{(s, t)\rho = (t^{-1}\varphi)s\}\{(s, t)\rho = s(t\varphi)\}$ $(s \in S, t \in T)$. The transformation group (S, T, ρ) is called the {*left*}{*right*} *transformation group of* S *induced by* T *under* φ.

1.59. REMARK. Let φ be a continuous homomorphism of the topological group T into the topological group S. Then the {left}{right} transformation group of S induced by T under φ coincides with the (T, φ)-restriction of the {left}{right} transformation group of S.

1.60. Remark. Let φ be a continuous homomorphism of a topological group T into a topological group S. Then the {left}{right} transformation group of S induced by T under φ is uniformly equicontinuous relative to the {left}{right} uniformity of S.

1.61. Remark. Let φ be a homomorphism of $\mathcal{I}$ into a topological group T, let $t = 1\varphi$ and let θ be the {left}{right} translation of T induced by $\{t^{-1}\}\{t\}$. Then the {left}{right} transformation group of T induced by $\mathcal{I}$ under φ coincides with the discrete flow on T generated by θ.

1.62. Definition. Let T be a topological group, let Y be a uniform space, let Φ be the class of all {right}{left} uniformly continuous functions on T to Y, let Φ be provided with its space-index uniformity and let $\rho : \Phi \times T \rightarrow \Phi$ be defined by $\{(\varphi, t)\rho = (t\tau\varphi \mid \tau \in T)\}\{(\varphi, t)\rho = (\tau t^{-1}\varphi \mid \tau \in T)\}$ $(\varphi \in \Phi, t \in T)$. The uniformly equicontinuous transformation group (Φ, T, ρ) is called the *{left}{right} uniform functional transformation group over T to Y*.

1.63. Definition. Let T be a locally compact topological group, let Y be a uniform space, let Φ be the class of all continuous functions on T to Y, let Φ be provided with its compact-index uniformity and let $\rho : \Phi \times T \rightarrow \Phi$ be defined by $\{(\varphi, t)\rho = (t\tau\varphi \mid \tau \in T)\}\{(\varphi, t)\rho = (\tau t^{-1}\varphi \mid \tau \in T)\}$ $(\varphi \in \Phi, t \in T)$. The transformation group (Φ, T, ρ) is called the *{left}{right} functional transformation group over T to Y*.

1.64. Remark. A particular case of 1.63 arises when T is discrete. In such an event a different notation may be used, as indicated by the following statements:

(1) $\Phi = Y^T = \times_{\tau \in T} Y_\tau$ where $Y_\tau = Y$ $(\tau \in T)$.

(2) The point-index (= compact-index) uniformity of Φ coincides with the product uniformity of $\times_{\tau \in T} Y_\tau$.

(3) If $y = (y_\tau \mid \tau \in T) \in \times_{\tau \in T} Y_\tau$ and if $t \in T$, then $\{(y, t)\rho = (y_{t\tau} \mid \tau \in T)\}$ $\{(y, t)\rho = (y_{\tau t^{-1}} \mid \tau \in T)\}$.

1.65. Lemma. *Let T be a locally compact topological group, let Y be a uniform space, let (Φ, T, ρ) be the {left}{right} functional transformation group over T to Y, let $\varphi \in \Phi$ and let $\Psi \subset \Phi$. Then:*

(1) *The orbit φT of φ is totally bounded if and only if φ is {left}{right} uniformly continuous and bounded.*

(2) *If Y is complete, then the orbit-closure $\overline{\varphi T}$ of φ is compact if and only if φ is {left}{right} uniformly continuous and bounded.*

(3) *ΨT is totally bounded if and only if Ψ is {left}{right} uniformly equicontinuous and bounded.*

(4) *If Y is complete, then $\overline{\Psi T}$ is compact if and only if Ψ is {left}{right} uniformly equicontinuous and bounded.*

Proof. Use 11.31 and 11.32.

1.66. Definition. Let X be a uniform space, let each transition $\pi^t : X \to X$ $(t \in T)$ be uniformly continuous, let the motion space $[\pi_x : T \to X \mid x \in X]$ be equicontinuous, let Y be a uniform space, let Φ be the class of all uniformly continuous functions on X to Y, let Φ be provided with its space-index uniformity and let $\rho : \Phi \times T \to \Phi$ be defined by $(\varphi, t)\rho = \pi^{t^{-1}}\varphi$ $(\varphi \in \Phi, t \in T)$. The uniformly equicontinuous transformation group (Φ, T, ρ) is called the *uniform functional transformation group over* (X, T, π) *to* Y.

1.67. Remark. Let T be a topological group, let (T, T, η) be the {left} {right} transformation group of T and let Y be a uniform space. Then the uniform functional transformation group over (T, T, η) to Y coincides with the {left}{right} uniform functional transformation group over T to Y.

1.68. Definition. Let T be locally compact, let Y be a uniform space, let Φ be the class of all continuous functions on X to Y, let Φ be provided with its compact-index uniformity and let $\rho : \Phi \times T \to \Phi$ be defined by $(\varphi, t)\rho = \pi^{t^{-1}}\varphi$ $(\varphi \in \Phi, t \in T)$. The transformation group (Φ, T, ρ) is called the *functional transformation group over* (X, T, π) *to* Y.

1.69. Remark. Let T be a locally compact topological group, let (T, T, η) be the {left}{right} transformation group of T and let Y be a uniform space. Then the functional transformation group over (T, T, η) to Y coincides with the {left}{right} functional transformation group over T to Y.

1.70. Notes and references.

(1.01) The concept of a transformation group for which the topology of the group plays a role appears to have originated in the latter part of the nineteenth century (cf., e.g., Lie and Engel [1]). A system of n differential equations of the first order defines, under suitable conditions, a transformation group (X, T, π) for which X is an n-dimensional manifold and T is the additive group of reals. Thus a classical dynamical system with n degrees of freedom defines a transformation group for which the *phase space* is the $2n$-dimensional manifold customarily associated with the term. See also Zippin [1].

(1.35) For a decomposition of a compact metric space, the equivalence of {star-open}{star-closed} with {lower semi-continuous}{upper semi-continuous} is readily verified (cf. Whyburn [1], Ch. VII).

(1.40) Φ-orbit partition transformation groups arise naturally in the study of geodesic flows on manifolds (cf. §13).

2. ORBIT-CLOSURE PARTITIONS

2.01. STANDING NOTATION. Throughout this section T denotes a topological group.

2.02. DEFINITION. A subset A of T is said to be $\{$*left*$\}$ $\{$*right*$\}$ *syndetic in* T provided that $\{T = AK\}$ $\{T = KA\}$ for some compact subset K of T.

2.03. REMARK. The following statements are valid.

(1) If $A \subset T$, then A is $\{$left$\}$ $\{$right$\}$ syndetic in T if and only if there exists a compact subset K of T such that every $\{$left$\}$ $\{$right$\}$ translate of K intersects A.

(2) If $A \subset B \subset T$ and if A is $\{$left$\}$ $\{$right$\}$ syndetic in T, then so also is B.

(3) If $A \subset T$, then A is $\{$left$\}$ $\{$right$\}$ syndetic in T if and only if A^{-1} is $\{$right$\}$ $\{$left$\}$ syndetic in T.

(4) If $A \subset T$ and if A is symmetric or invariant (in particular, if A is a subgroup of T or if T is abelian), then A is left syndetic in T if and only if A is right syndetic in T. In such an event, the equivalent phrases "left syndetic", "right syndetic" are contracted to "syndetic".

(5) If A is a syndetic subgroup of T, then the left, right quotient spaces T/A, $T\backslash A$ are compact.

(6) If T is locally compact, if A is a subgroup of T and if some one of the left, right quotient spaces T/A, $T\backslash A$ is compact, then A is syndetic in T.

(7) If T is discrete and if A is a subgroup of T, then A is syndetic in T if and only if A is of finite index in T.

(8) If A is a $\{$left$\}$ $\{$right$\}$ syndetic subset of T and if U is a neighborhood of e, then $\{AU\}$ $\{UA\}$ is $\{$left$\}$ $\{$right$\}$ syndetic relative to the discrete topology of T.

2.04. EXAMPLE. Let T be the discrete free group on 2 generators a, b and let $\{A\}$ $\{B\}$ be the set of all words of T which in reduced form do not $\{$end$\}$ $\{$begin$\}$ with $\{a^1\}$ $\{b^1\}$. Then:

(1) A is left syndetic in T but A is not right syndetic in T.

(2) B is right syndetic in T but B is not left syndetic in T.

(3) $A \cup B$ is both left and right syndetic in T but there is no compact (= finite) subset K of T such that every bilateral translate of K intersects $A \cup B$.

2.05. DEFINITION. Let G be a group. A *semigroup in* G is defined to be a subset H of G such that $HH \subset H$.

2.06. LEMMA. *Let S be a left or right syndetic closed semigroup in T. Then S is a subgroup of T.*

PROOF. We assume without loss that S is left syndetic. Let $s \in S$ and let

U be a neighborhood of the identity e of T. It is enough to show that $s^{-1}U \cap S \neq \emptyset$. Let V be a neighborhood of e such that $VV^{-1} \subset U$ and let K be a compact subset of T such that $T = SK$. There exists a finite class $\mathfrak{F}$ of right translates of V such that $K \subset \bigcup \mathfrak{F}$. Choose $k_0 \in K$. Now $s^{-1}k_0 = s_1k_1$ for some $s_1 \in S$ and some $k_1 \in K$. Again $s^{-1}k_1 = s_2k_2$ for some $s_2 \in S$ and some $k_2 \in K$. This may be continued. Thus there exist sequences $k_0, k_1, \cdots$ in K and $s_1, s_2, \cdots$ in S such that $s^{-1}k_i = s_{i+1}k_{i+1} (i = 0, 1, \cdots)$. Select integers $m, n (0 \leqq m < n)$ and $V_0 \in \mathfrak{F}$ such that $k_m, k_n \in V_0$. Now $s^{-1}k_mk_n^{-1} = (s^{-1}k_mk_{m+1}^{-1})(k_{m+1}k_{m+2}^{-1}) \cdots (k_{n-1}k_n^{-1}) = s_{m+1}ss_{m+2} \cdots ss_n \in S$. Also $s^{-1}k_mk_n^{-1} \in s^{-1}V_0V_0^{-1} \subset s^{-1}U$. Hence $s^{-1}U \cap S \neq \emptyset$. The proof is completed.

2.07. Standing notation. For the remainder of this section (X, T, π) denotes a transformation group.

2.08. Definition. Let $x \in X$ and let $S \subset T$. The *x-envelope of S*, denoted S_x, is defined to be the subset $[t \mid t \in T, xt \in \overline{xS}]$ of T.

2.09. Lemma. *Let $x \in X$. Then:*

(1) *If $S \subset T$, then $S_x = \overline{xS}\pi_x^{-1}$, S_x is closed in T, $S_x \supset S$ and $\overline{xS_x} = \overline{xS}$.*

(2) *If S is an invariant semigroup in T, then S_x is a semigroup in T.*

Proof. (1) Obvious.

(2) $xS_xS_x \subset \overline{xS}S_x \subset \overline{xSS_x} = \overline{xS_xS} \subset \overline{\overline{xS}S} = \overline{xSS} \subset \overline{xS}$ whence $S_xS_x \subset S_x$.

2.10. Lemma. *Let $x \in X$ and let S be a syndetic invariant subgroup of T. Then:*

(1) *S_x is a subgroup of T.*

(2) *If U is a neighborhood of e, then $x \notin \overline{x(T - S_xU)}$.*

(3) *If T is locally compact and if U is a neighborhood of x, then there exists a compact subset M of T such that $xM \subset U$ and $S_x \subset SM^{-1}$.*

Proof. (1) Use 2.06 and 2.09.

(2) We first show that if $t \in T - S_x$, then $x \notin \overline{xS_xV_0}$ for some neighborhood V_0 of t. Let $t \in T - S_x$. Since $t^{-1} \notin S_x$ by (1), $xt^{-1} \notin \overline{xS_x}$ and $x \notin \overline{xS_xt}$. There exist neighborhoods W of x and V of e such that $V = V^{-1}$ and $WV \cap xS_xt = \emptyset$. It follows that $W \cap xS_xtV = \emptyset$. Define $V_0 = tV$.

We may assume that U is open. Let K be a compact subset of T such that $T = SK$. Define $H = K - S_xU$. Using (1) we conclude that $T = SK \subset S(H \cup S_xU) \subset SH \cup SS_xU \subset S_xH \cup S_xU$ and $S_xH \cap S_xU = \emptyset$. Hence $T - S_xU = S_xH$. By the preceding paragraph to each $t \in H$ there corresponds a neighborhood V_t of t such that $x \notin \overline{xS_xV_t}$. Since H is compact, there exists a finite subset E of H such that $H \subset \bigcup_{t \in E} V_t$. Hence $x \notin \overline{xS_xH} = \overline{x(T - S_xU)}$.

(3) We may assume that U is open. Let K be a compact subset of T such that $T = SK$. Define $H = K \cap S_x$. If $t \in H$, then $xt \in \overline{xS}$ and $x \in \overline{xSt^{-1}}$. Then $t \in H$ implies the existence of $s_t \in S$ such that $xs_tt^{-1} \in U$ and hence the existence of a compact neighborhood V_t of s_tt^{-1} such that $xV_t \subset U$. Since H is compact, there is a finite subset E of H for which $H \subset \bigcup_{t \in E} V_t^{-1}s_t$. Define $M = \bigcup_{t \in E} V_t$. Clearly $xM \subset U$. If $t \in S_x$, then $t = sk$ for some $s \in S$ and

some $k \in K$, $k = s^{-1}t \in S_x$ and $t \in S(K \cap S_x) = SH$. Thus $S_x \subset SH$. Since $SH \subset \bigcup_{t \in E} V_t^{-1}S = SM^{-1}$, we have $S_x \subset SM^{-1}$. The proof is completed.

2.11. Definition. Let $A \subset X$ and let $S \subset T$. The set A is said to be *minimal under* S or S-*minimal* provided that A is an orbit-closure under S and A does not contain properly an orbit-closure under S. When $S = T$, the phrase "under T" and the prefix "T-" may be omitted. We often use the more colorful phrase "minimal orbit-closure" in preference to "minimal set".

2.12. Remark. Let $A \subset X$. Then the following statements are pairwise equivalent:

(1) A is a minimal orbit-closure under T.

(2) $A \neq \emptyset$ and $\overline{xT} = A$ for each $x \in A$.

(3) A is nonvacuous closed T-invariant and A is minimal with respect to this property.

(4) A is nonvacuous closed and $UT = A$ for each nonvacuous subset U of A which is open in A.

2.13. Remark. Let A be a minimal orbit-closure under T. Then:

(1) A is open in X if and only if int $A \neq \emptyset$.

(2) If int $A \neq \emptyset$, then A is a union of components of X.

(3) If int $A \neq \emptyset$, and if T is connected, then A is a component of X.

(4) If int $A \neq \emptyset$, and if X is connected, then $A = X$.

2.14. Lemma. *Let n be a positive integer, let X be an n-dimensional manifold and let $A \subset X$. Then* dim $A = n$ *if and only if* int $A \neq \emptyset$.

Proof. Cf. [Hurewicz-Wallman [1], pp. 44–46].

2.15. Theorem. *Let n be a positive integer, let X be an n-dimensional manifold and let A be a minimal orbit-closure under T such that $A \neq X$. Then* dim $A \leqq n - 1$.

Proof. Use 2.13 (4) and 2.14.

2.16. Definition. A *Cantor-manifold* is defined to be a compact metrizable space X of positive finite dimension n such that X is not disconnected by a subset of dimension $\leqq n - 2$.

2.17. Lemma. *Let X be a compact metrizable space of positive finite dimension. Then there exists a subset C of X such that C is a Cantor-manifold and* dim $C =$ dim X.

Proof. Cf. [Hurewicz-Wallman [1], pp. 94–95].

2.18. Theorem. *Let X be a finite-dimensional compact metrizable space such that X contains more than one point, let X be minimal under T and let T be connected. Then X is a Cantor-manifold and hence X has the same dimension at every point of X.*

PROOF. Let $n = \dim X$. Assume X is not a Cantor-manifold. Then there exist closed proper subsets A, B of X such that $X = A \cup B$ and $\dim(A \cap B) \leqq n - 2$ (cf. [Hurewicz-Wallman] [1], p. 94). By 2.17 there exists $C \subset X$ such that C is a Cantor-manifold and $\dim C = n$. Let $\{E\}$ $\{F\}$ be the set of all $t \in T$ such that $\{Ct \subset A\}$ $\{Ct \subset B\}$. Clearly E and F are closed disjoint subsets of T. If $t \in T$, then Ct is an n-dimensional Cantor-manifold and hence $Ct \subset A$ or $Ct \subset B$. Thus $T = E \cup F$. If follows that $T = E$ or $T = F$. If $\{T = E\}$ $\{T = F\}$, then $\{CT \subset A\}$ $\{CT \subset B\}$ and $\overline{CT} \neq X$. This contradicts the minimality of X. The proof is completed.

2.19. REMARK. If the hypothesis that T be connected is omitted from 2.18, the second conclusion fails. Cf. [Floyd [1]].

2.20. REMARK. Let φ be a continuous homomorphism of a topological group T into a topological group S and let (S, T, ρ) be the {left} {right} transformation group of S induced by T under φ. Then S is minimal under (S, T, ρ) if and only if $\overline{T\varphi} = S$.

2.21. REMARK. If A and B are minimal orbit-closures under T, then $A \cap B = \emptyset$ or $A = B$. In other words, the class of all minimal orbit-closures under T is disjoint.

2.22. THEOREM. *Let X be compact and nonvacuous. Then there exists a minimal orbit-closure under T.*

PROOF. Let $\mathfrak{A}$ be the class of all nonvacuous closed invariant subsets of X. Since $X \in \mathfrak{A}$, we have $\mathfrak{A} \neq \emptyset$. Partially order $\mathfrak{A}$ by inclusion. By the extremum law there exists a minimal element A of $\mathfrak{A}$. By 2.12, A is a minimal orbit-closure under T. The proof is completed.

2.23. REMARK. The following statements are pairwise equivalent:
(1) The class of all orbit-closures under T is a partition of X.
(2) If $x \in X$ and if $y \in \overline{xT}$, then $x \in \overline{yT}$.
(3) Every orbit-closure under T is minimal under T.
(4) The class of all minimal orbit-closures under T is a covering of X.
(5) The class of all minimal orbit-closures under T is a partition of X.

2.24. INHERITANCE THEOREM. *Let S be a syndetic invariant subgroup of T. Then the class of all orbit-closures under S is a partition of X if and only if the class of all orbit-closures under T is a partition of X.*

PROOF. Assume that the class of all orbit-closures under T is a partition of X. Let $x \in X$ and $y \in \overline{xS}$. It is enough to show that $\overline{yS} = \overline{xS}$. Let K be a compact subset of T for which $T = SK$. Since $x \in \overline{yT} = \overline{ySK^{-1}} = \overline{yS}K^{-1}$ by 1.18 (6), there exists $k \in K$ such that $xk \in \overline{yS}$. It follows that $\overline{xkS} \subset \overline{yS} \subset \overline{xS}$. Since S_x is a subgroup of T by 2.10 (1) and $k \in S_x$, we have $\overline{xS} = \overline{xS_x} = \overline{xS_x k} = \overline{xS_x}k = \overline{xS}k = \overline{xSk} = \overline{xkS} \subset \overline{yS} \subset \overline{xS}$. Thus $\overline{yS} = \overline{xS}$.

Assume that the class of all orbit-closures under S is a partition of X. Let $x \in X$ and $y \in \overline{xT}$. It is enough to show that $x \in \overline{yT}$. Let K be a compact subset of T for which $T = SK$. Since $y \in \overline{xT} = \overline{xSK^{-1}} = \overline{xS}K^{-1}$, there exists $k \in K$ such that $yk \in \overline{xS}$. Now $x \in \overline{ykS}$. Hence $x \in \overline{yT}$. The proof is completed.

2.25. THEOREM. *Let X be minimal under T, let S be an invariant subgroup of T, let K be a compact subset of T, let $T = SK$ and let $\mathfrak{A}$ be the class of all orbit-closures under S. Then $\mathfrak{A}$ is a partition of X and*

$$\text{crd } \mathfrak{A} = \text{crd } T\backslash S_x \leqq \text{crd } K(x \in X).$$

PROOF. By 2.24, $\mathfrak{A}$ is a partition of X. If $x \in X$ and if $\tau, \sigma \in T$, then $\overline{x\tau S} = \overline{x\sigma S}$ if and only if $\tau\sigma^{-1} \in S_x$. The conclusion follows.

2.26. REMARK. Let X be minimal under T, let $x, y \in X$, let S be a syndetic invariant subgroup of T and for $t \in T$ define $\varphi_t : S \to S$ by $s\varphi_t = t^{-1}st$ $(s \in S)$. Then there exists $t \in T$ such that (π^t, φ_t) is a topological isomorphism of the transformation group $(\overline{xS}, S, \pi \mid \overline{xS} \times S)$ onto the transformation group $(\overline{yS}, S, \pi \mid \overline{yS} \times S)$.

PROOF. Let K be a compact subset of T such that $T = SK$. Since $X = \overline{xT} = \overline{xS}K$, there exists $t \in K$ such that $y \in \overline{xS}t = \overline{xtS}$ whence $\overline{xS}t = \overline{yS}$ by 2.24. The conclusion follows.

2.27. DEFINITION. A subset A of X is said to be *totally minimal under T* provided that A is a minimal orbit-closure under every syndetic invariant subgroup of T.

2.28. THEOREM. *Let T be discrete. Then every connected minimal orbit-closure under T is totally minimal under T.*

PROOF. Use 2.25.

2.29. REMARK. Let $\mathfrak{A}$ be the class of all orbits under T. Then:
(1) If $E \subset X$, then $ET = E\mathfrak{A}$.
(2) $\mathfrak{A}$ is star-open.

2.30. REMARK. Let the class $\mathfrak{A}$ of all orbit-closures under T be a partition of X. Then:
(1) If $E \subset X$, then $ET \subset E\mathfrak{A} = \bigcup_{x \in E} \overline{xT} \subset \overline{ET}$.
(2) If E is an open subset of X, then $E\mathfrak{A} = ET$.
(3) $\mathfrak{A}$ is star-open.

2.31. REMARK. Consider the following statements:
(I) The class of all orbit-closures under T is a star-closed partition of X.
(II) If $x \in X$ and if U is a neighborhood of $\overline{xT}$, then there exists a neighborhood V of x such that $y \in V$ implies $\overline{yT} \subset U$.
(III) If $x \in X$ and if U is a neighborhood of $\overline{xT}$, then there exists a neighborhood V of x such that $VT \subset U$.
(IV) If $x \in X$ and if $\mathfrak{N}$ is the neighborhood filter of x, then $\bigcap_{V \in \mathfrak{N}} \overline{VT} = \overline{xT}$.
Then:
(1) If X is a T_1-space, then I is equivalent to II.

(2) If X is a regular space and if every orbit-closure under T is compact, then I is equivalent to III.

(3) If X is a compact T_2-space, then I is equivalent to IV.

2.32. **Inheritance Theorem.** *Let X be compact, let X be minimal under T and let S be a syndetic invariant subgroup of T. Then the class of all orbit-closures under S is a star-closed decomposition of X.*

Proof. Let $\mathfrak{A}$ be the class of all orbit-closures under S. By 2.24, $\mathfrak{A}$ is a decomposition of X. Let $x \in X$ and let K be a compact subset of T for which $T = SK$. Since $X = \overline{xT} = \overline{xSK} = x\overline{SK}$, it follows that $\mathfrak{A} = [Ck \mid k \in K]$ where $C = \overline{xS}$. Let E be a closed subset of X. Define $H = [t \mid t \in T, Ct \cap E \neq \emptyset]$. By 1.18 (4), $T - H$ is open whence H is closed in T. Define $M = H \cap K$. Now M is compact and therefore $CM = E\mathfrak{A}$ is closed in X. The proof is completed.

2.33. **Theorem.** *Let X be a locally connected {normal T_1-space}{T_2-space}, let T be discrete, let the class of all orbit-closures under T be a star-closed {partition}{decomposition} of X and let S be a syndetic invariant subgroup of T. Then the class of all orbit-closures under S is a star-closed {partition}{decomposition} of X.*

Proof. By 2.24, the class of all orbit-closures under S is a {partition}{decomposition} of X. Let $x \in X$ and let U be a neighborhood of $\overline{xS}$. There exists a finite subset K of T such that $T = SK$. Since $\overline{xT} = \overline{xSK} = \overline{xKS} = \bigcup_{t \in K} \overline{xtS}$, there exists a finite subset H of T such that $e \in H$, $\overline{xT} = \bigcup_{t \in H} \overline{xtS}$, and $\overline{xtS} \cap \overline{xsS} = \emptyset$ $(t, s \in H; t \neq s)$. For each $t \in H$ choose an open neighborhood U_t of $\overline{xtS}$ so that $U_e \subset U$ and $U_t \cap U_s = \emptyset$ $(t, s \in H; t \neq s)$. There exists a connected neighborhood V of x such that $y \in V$ implies $\overline{yT} \subset \bigcup_{t \in H} U_t$. It remains to show that $y \in V$ implies $\overline{yS} \subset U$. Assume there exists $y \in V$ such that $\overline{yS} \not\subset U$. Then $ys \in U_t$ for some $s \in S$ and for some $t \in H$ with $t \neq e$. It follows that the connected set Vs intersects U_e and U_t whence $Vs \cap \operatorname{bdy} U_e \neq \emptyset$. This contradicts $VT \subset \bigcup_{t \in H} U_t$. The proof is completed.

2.34. **Definition.** Let $(X, \mathfrak{U})$ be a uniform space and let $\mathfrak{A}$ be a partition of X. For $\alpha \in \mathfrak{U}$ let $\alpha^* = [(A, B) \mid A, B \in \mathfrak{A}, A \subset B\alpha, B \subset A\alpha]$. The uniformity of $\mathfrak{A}$ generated by the uniformity base $[\alpha^* \mid \alpha \in \mathfrak{U}]$ of $\mathfrak{A}$ is called the *partition uniformity of* $\mathfrak{A}$. The partition uniformity of $\mathfrak{A}$ may fail to be separated.

2.35. **Definition.** Let X be a uniform space and let $\mathfrak{A}$ be a partition of X. The partition $\mathfrak{A}$ is said to be *star-indexed* provided that the following equivalent conditions are satisfied: (1–2) If α is an index of X, then there exists an index β of X such that $x \in X$ implies $\{x\mathfrak{A}\beta \subset x\alpha\mathfrak{A}\}\{x\beta\mathfrak{A} \subset x\mathfrak{A}\alpha\}$.

2.36. **Remark.** Let $(X, \mathfrak{U})$ be a uniform space, let $\mathfrak{A}$ be a star-indexed partition of X, let $\mathfrak{V}$ be the partition uniformity of $\mathfrak{A}$ and for $\alpha \in \mathfrak{U}$ let $\hat{\alpha} = [(x\mathfrak{A}, y\mathfrak{A}) \mid (x, y) \in \alpha]$. Then:

(1) $[\hat{\alpha} \mid \alpha \in \mathfrak{U}]$ is a base of $\mathfrak{V}$.

(2) The projection of $(X, \mathfrak{U})$ onto $(\mathfrak{A}, \mathfrak{V})$ is uniformly continuous and uniformly open.

(3) $\mathfrak{A}$ is star-open.

(4) The topology of $\mathfrak{A}$ induced by $\mathfrak{V}$ coincides with the partition topology of $\mathfrak{A}$.

2.37. REMARK. Let X be a uniform space and let $\mathfrak{A}$ be a decomposition of X. Then:

(1) If $\mathfrak{A}$ is star-indexed, then $\mathfrak{A}$ is star-open and star-closed.

(2) If X is compact and if $\mathfrak{A}$ is star-open and star-closed, then $\mathfrak{A}$ is star-indexed.

2.38. THEOREM. *Let X be a uniform space, let T be equicontinuous and let $\mathfrak{A}$ be the class of all orbit-closures under T. Then:*

(1) *$\mathfrak{A}$ is a partition of X.*

(2) *If every orbit-closure under T is compact, then $\mathfrak{A}$ is a star-closed decomposition of X.*

PROOF. (1) Let $x, y \in X$ and let $y \in \overline{xT}$. It is enough to show that $x \in \overline{yT}$. Let α be a symmetric index of X. There exists a neighborhood U of y such that $Ut \subset yt\alpha$ $(t \in T)$. Choose $t \in T$ such that $xt \in U$. Then $x \in Ut^{-1} \subset yt^{-1}\alpha$, $yt^{-1} \in x\alpha$ and $yT \cap x\alpha \neq \emptyset$. This shows $x \in \overline{yT}$.

(2) Let $A \in \mathfrak{A}$ and let α be an index of X. For each $x \in X$ there exists a neighborhood U_x of x such that $U_x t \subset xt\alpha$ $(t \in T)$. Define $U = \bigcup_{x \in A} U_x$. Then U is a neighborhood of A and $UT \subset xT\alpha$. The proof is completed.

2.39. THEOREM. *Let X be a uniform space, let T be uniformly equicontinuous and let $\mathfrak{A}$ be the class of all {orbits} {orbit-closures} under T. Then $\mathfrak{A}$ is star-indexed.*

PROOF. Obvious.

2.40. DEFINITION. Let $x \in X$ and let $\mathfrak{S}$ be the class of all closed syndetic invariant subgroups of T. The *trace of x under T* or the *T-trace of x* is defined to be the subset $\bigcap_{S \in \mathfrak{S}} \overline{xS}$ of X. A *trace under T* or a *T-trace* is defined to be a subset A of X such that A is the T-trace of some point of X. The phrase "under T" and the prefix "T-" may be omitted when there is no chance for ambiguity.

2.41. REMARK. Let $\mathfrak{S}$ be the class of all closed syndetic invariant subgroups of T and for $x \in X$ let $x^* = \bigcap_{S \in \mathfrak{S}} \overline{xS}$ be the T-trace of x. Then:

(1) If $x \in X$, then $x \in x^*$.

(2) If $x, y \in X$, then $x^* \subset y^*$ if and only if $x \in y^*$.

(3) If $x, y \in X$, then $x^* = y^*$ if and only if $x \in y^*$ and $y \in x^*$.

(4) $[x^* \mid x \in X]$ is a covering of X.

(5) $[x^* \mid x \in X]$ is a partition of X if and only if $x, y \in X$ with $x \in y^*$ implies $x^* = y^*$.

2.42. REMARK. Let $x, y \in X$. Then the following statements are equivalent:

(1) The T-traces of x and y coincide.

(2) If S is a closed syndetic invariant subgroup of T, then $\overline{xS} = \overline{yS}$.

2.43. Remark. Let X be a compact uniform space, let X be minimal under T and let $x, y \in X$. Then the following statements are equivalent:

(1) The T-traces of x and y coincide.

(2) If α is an index of X and if S is a closed syndetic invariant subgroup of T, then there exists $s, r \in S$ such that $(xs, yr) \in \alpha$.

2.44. Notes and references.

(2.02) The term syndetic is from the Greek συνδετικός, meaning *to bind together*. When $T = \mathcal{R}$, a subset of T is syndetic if and only if it is relatively dense.

(2.06) This lemma is proved in Gottschalk [8].

(2.11) The concept of a minimal set is due to G. D. Birkhoff (cf. Birkhoff [1], vol. 1, pp. 654–672).

(2.15) A restricted form of this theorem is proved in Hilmy [1].

(2.18) The proof is essentially that due to A. Markoff for the case of a continuous flow (cf. Markoff [1]).

(2.22) Concerning the *extremum law*, see Gottschalk [11]. A proof of the existence of a minimal set for the case of a flow in a compact space was given by Birkhoff (see (2.11)).

(2.40) The concept of T-trace appears in the thesis of Gottschalk [1].

3. RECURSION

3.01. Standing notation. Throughout this section (X, T, π) denotes a transformation group. We shall often use the phrase "the transformation group T" or simply the symbol "T" to stand for the phrase "the transformation group (X, T, π)."

3.02. Definition. Let $x \in X$. The *period of T at x* or the *period of x under T* is defined to be the greatest subset P of T such that $xP = x$.

3.03. Remark. Let $x \in X$ and let P be the period of T at x. Then

(1) $P = x\pi_x^{-1}$.

(2) P is a subgroup of T.

(3) If X is a T_0-space, then P is closed in T.

(4) If $t \in T$, then $t^{-1}Pt$ is the period of T at xt.

Proof. (3) Let $t \in \overline{P}$. Since $x\overline{P} \subset \overline{xP} = \bar{x}$, we have $xt \in \bar{x}$ and $\overline{xt} \subset \bar{x}$. Now $t^{-1} \in \overline{P}^{-1} = \overline{P^{-1}} = \overline{P}$. As before, $xt^{-1} \in \bar{x}$ whence $x \in \bar{x}t = \overline{xt}$ and $\bar{x} \subset \overline{xt}$. We conclude that $\overline{xt} = \bar{x}$, $xt = x$ and $t \in P$.

3.04. Definition. The *period of T* is defined to be the greatest subset P of T such that $x \in X$ implies $xP = x$.

3.05. Remark. Let P be the period of T. Then:

(1) P is an invariant subgroup of T.

(2) If P_x $(x \in X)$ is the period of T at x, then $P = \bigcap_{x \in X} P_x$.

(3) T is effective if and only if $P = [e]$.

(4) If X is a T_0-space, then P is closed in T and the transformation group $(X, T/P, \rho)$ is effective where $\rho : X \times T/P \to X$ is defined by $(x, tP)\rho = xt$ $(x \in X, t \in T)$.

3.06. Definition. Let $x \in X$. The transformation group T is said to be *periodic at x* and the point x is said to be *periodic under T* provided that the period of T at x is a syndetic subset of T. The transformation group T is said to be *fixed at x* and the point x is said to be *fixed under T* provided that $xT = x$, that is, the period of T at x is T.

The transformation group T is said to be *pointwise periodic* provided that T is periodic at every point of X.

The transformation group T is said to be *periodic* provided that the period of T is a syndetic subset of T.

3.07. Remark. Let $x \in X$ and let T be periodic at x. Then:

(1) T is pointwise periodic on xT.

(2) If T is abelian, then T is periodic on xT.

(3) xT is compact.

3.08. Lemma. *Let $x \in X$, let P be the period of x under T and let $\varphi : T\backslash P \to xT$ be defined by $(Pt)\varphi = xt$ $(t \in T)$. Then:*

(1) *$\varphi : T\backslash P \to xT$ is one-to-one continuous onto.*

(2) *The following statements are pairwise equivalent:*

(I) *$\varphi : T\backslash P \to xT$ is homeomorphic.*

(II) *$\pi_x : T \to xT$ is open.*

(III) *$\pi_x : T \to xT$ is open at e.*

(3) *If xT is a second-category T_1-space and if T is separable locally compact, then $\varphi : T\backslash P \to xT$ is homeomorphic.*

Proof. (3) We may assume that $X = xT$.

We first show that if V is a neighborhood of e, then int $(xV) \neq \emptyset$. Let V be a neighborhood of e. Choose a compact neighborhood W of e such that $W \subset V$. There exists a countable subset C of T such that $T = WC$. Now $X = xT = xWC = \bigcup_{t \in C} xWt$ and each set xWt $(t \in C)$ is closed in X. If int $(xW) = \emptyset$, then int $(xWt) = \emptyset$ $(t \in C)$ and X is of the first category. Hence int $(xW) \neq \emptyset$. Since $xW \subset xV$, we have int $(xV) \neq \emptyset$.

It is enough by (2) to show that $\pi_x : T \to X$ is open at e. Let U be a neighborhood of e. Choose a neighborhood V of e such that $VV^{-1} \subset U$. Now int $(xV) \neq \emptyset$. Hence there exists $t \in V$ such that xV is a neighborhood of xt. It follows that xVt^{-1} is a neighborhood of x, xVV^{-1} is a neighborhood of x and $xU = U\pi_x$ is a neighborhood of x. The proof is completed.

3.09. Theorem. *Let $x \in X$, let X be a T_2-space and let T be separable locally compact. Then x is periodic under T if and only if xT is compact.*

Proof. Use 3.08 (3).

3.10. Remark. The following statements are valid:

(1–2) If $x \in X$, then T is periodic at x if and only if there exists a syndetic {subset}{subgroup} A of T such that $xA = x$.

(3–4) T is periodic if and only if there exists a syndetic {subset}{invariant subgroup} A of T such that $x \in X$ implies $xA = x$.

3.11. Remark. The recursion notions defined below are generalizations of periodicity notions.

3.12. Standing notation. Let there be distinguished in the phase group T certain subsets which are called *admissible*. Let $\mathfrak{A}$ denote the class of all admissible subsets of T.

3.13. Definition. Let $x \in X$. The transformation group T is said to be *recursive at* x and the point x is said to be *recursive under* T or *T-recursive* provided that if U is a neighborhood of x, then there exists an admissible subset A of T such that $xA \subset U$.

The transformation group T is said to be *pointwise recursive* provided that T is recursive at every point of X.

Let X be a uniform space. The transformation group T is said to be *recursive* provided that if α is an index of X, then there exists an admissible subset A of T such that $x \in X$ implies $xA \subset x\alpha$.

Let $x \in X$. The transformation group T is said to be *locally recursive at* x and the point x is said to be *locally recursive under* T or *locally* T*-recursive* provided that if U is a neighborhood of x, then there exist a neighborhood V of x and an admissible subset A of T such that $VA \subset U$.

The transformation group T is said to be *locally recursive* provided that T is locally recursive at every point of X.

Let X be a uniform space. The transformation group T is said to be *weakly*$_1$ *recursive* provided that if α is an index of X, then there exist an admissible subset A of T and a compact subset K of T such that $x \in X$ implies the existence of an admissible subset B of T such that $A \subset BK$ and $xB \subset x\alpha$.

Let X be a uniform space. The transformation group T is said to be *weakly*$_2$ *recursive* provided that if α is an index of X, then there exist an admissible subset A of T and a finite subset K of T such that $x \in X$ implies the existence of $k, h \in K$ such that $xkAh \subset x\alpha$.

Let $x \in X$. The transformation group T is said to be *locally weakly recursive at* x and the point x is said to be *locally weakly recursive under* T or *locally weakly* T*-recursive* provided that if U is a neighborhood of x, then there exist a neighborhood V of x, an admissible subset A of T and a compact subset K of T such that $y \in V$ implies the existence of an admissible subset B of T such that $A \subset BK$ and $yB \subset U$.

The transformation group T is said to be *locally weakly recursive* provided that T is locally weakly recursive at every point of X.

Let $x \in X$. The transformation group T is said to be *regionally recursive at* x and the point x is said to be *regionally recursive under* T or *regionally* T*-recursive* provided that if U is a neighborhood of x, then there exists an admissible subset A of T such that $a \in A$ implies $U \cap Ua \neq \emptyset$.

The transformation group T is said to be *regionally recursive* provided that T is regionally recursive at every point of X.

3.14. Definition. In agreement with 1.03 a recursion property of the transformation group T is said to be *intrinsic* provided that the property "admissible subset of T" can be characterized solely in terms of the topological and group structures of T.

3.15. Definition. An intrinsic recursion property relative to the discrete topology of T is indicated by placing the {adjective *discrete*} {adverb *discretely*} before the {substantive} {adjectival} phrase which refers to the property.

3.16. Remark. When an intrinsic recursion property is characterized intrinsically but independently of the topology of T, then this recursion property implies the corresponding discrete recursion property (and indeed the corresponding recursion property relative to every topology of T which makes (X, T, π) a transformation group).

3.17. Remark. The relative strength of recursion properties is indicated in Table 1. No reverse implication is universally valid. (See Part II.)

3.18. Remark. The following statements are valid:

(1) If T is recursive, then T is weakly$_2$ recursive.

(2) If T is weakly$_2$ recursive and if $t\mathcal{A}s \subset \mathcal{A}$ $(t, s \in T)$, then T is pointwise recursive.

3.19. Remark. Our intent is that weakly recursive should mean the strongest property such that, under reasonable hypotheses, the following statement holds: If $x \in X$ and if T is recursive at x, then T is weakly recursive on $\overline{xT}$. Sometimes weakly$_1$ recursive is indicated and sometimes weakly$_2$ recursive.

3.20. Remark. The following are certain "universal" theorems on recursion properties.

Table 1

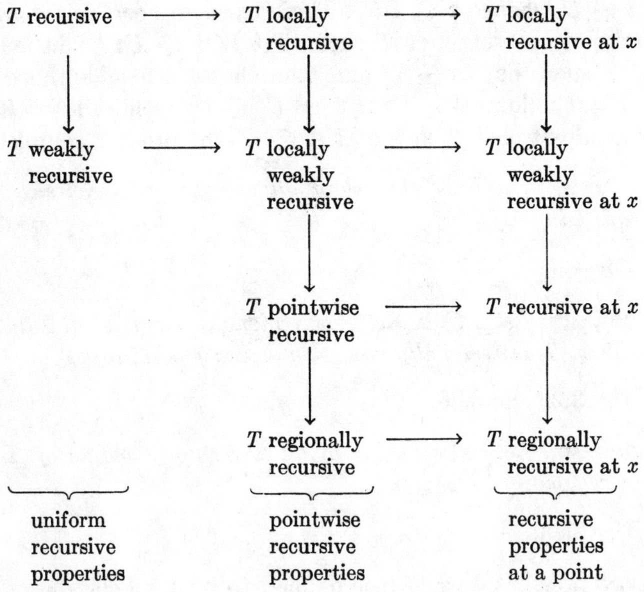

3.21. Theorem. *Let $x \in X$, let T be {recursive}{locally recursive}{locally weakly recursive}{regionally recursive} at x and let $t\mathcal{A}t^{-1} \subset \mathcal{A}$ $(t \in T)$. Then T is {recursive}{locally recursive}{locally weakly recursive}{regionally recursive} at every point of xT.*

Proof. Use 1.11.

3.22. Theorem. *Let R be the set of all {T-recursive}{locally T-recursive}{locally weakly T-recursive}{regionally T-recursive} points of X and let $t\mathcal{A}t^{-1} \subset \mathcal{A}$ $(t \in T)$. Then R is invariant under T.*

PROOF. Use 3.21.

3.23. THEOREM. *Let $x \in X$, let T be {recursive}{regionally recursive} at x and let φ be a continuous mapping of X into X such that $\pi^t\varphi = \varphi\pi^t$ $(t \in T)$. Then T is {recursive}{regionally recursive} at $x\varphi$.*

PROOF. Obvious.

3.24. THEOREM. *Let $x \in X$, let T be {locally recursive}{locally weakly recursive} at x and let φ be a continuous-open mapping of X onto X such that $\pi^t\varphi = \varphi\pi^t$ $(t \in T)$. Then T is {locally recursive}{locally weakly recursive} at $x\varphi$.*

PROOF. Obvious.

3.25. THEOREM. *Let $x \in X$, let T be locally recursive at x, let $\overline{xT}$ be minimal under T, and let $t\mathcal{A}s \subset \mathcal{A}$ $(t, s \in T)$. Then T is locally recursive at every point of $\overline{xT}$.*

PROOF. Let $y \in \overline{xT}$ and let U be an open neighborhood of y. Choose $t \in T$ such that $xt \in U$ whence $x \in Ut^{-1}$. There exist an open neighborhood W of x and an admissible subset B of T such that $WB \subset Ut^{-1}$ whence $WBt \subset U$. Choose $s \in T$ such that $ys \in W$ and then choose a neighborhood V of y such that $Vs \subset W$. It follows that $VsBt \subset WBt \subset U$. Define $A = sBt$. Since A is an admissible subset of T such that $VA \subset U$, the proof is completed.

3.26. THEOREM. *Let R be the set of all regionally T-recursive points of X. Then R is closed in X.*

PROOF. Obvious.

3.27. THEOREM. *Let $x \in X$, let T be regionally recursive at x and let $t\mathcal{A}t^{-1} \subset \mathcal{A}$ $(t \in T)$. Then T is regionally recursive at every point of $\overline{xT}$.*

PROOF. Use 3.22 and 3.26.

3.28. THEOREM. *Let R be the set of all T-recursive points of X. Then every point of $\overline{R}$ is regionally T-recursive.*

PROOF. Use 3.26.

3.29. DEFINITION. A *base of* $\mathcal{A}$ is defined to be a subclass $\mathcal{B}$ of $\mathcal{A}$ such that $A \in \mathcal{A}$ implies the existence of $B \in \mathcal{B}$ such that $B \subset A$.

3.30. THEOREM. *For $\alpha \subset X \times X$ and $A \subset T$ let $E(A, \alpha)$ denote the set of all $x \in X$ such that $xA \times xA \subset \alpha$. Let X be a uniform space, let $\mathcal{V}$ be a base of the uniformity of X, let $\mathcal{B}$ be a base of $\mathcal{A}$ such that $B \in \mathcal{B}$ implies $e \in B$ and let R be the set of all T-recursive points of X. Then $R = \bigcap_{\beta \in \mathcal{V}} \bigcup_{B \in \mathcal{B}} E(B, \beta)$.*

PROOF. Obvious.

3.31. THEOREM. *Let X be metrizable, let $P_1, P_2, \cdots$ be a sequence of subsets of T, for $A \subset T$ let $A \in \mathcal{A}$ if and only if $A \cap P_n \neq \emptyset$ for every positive integer n, and let R be the set of all T-recursive points of X. Then:*

(1) *R is a G_δ subset of X.*

(2) *If T is regionally recursive, then R is a residual subset of X.*

PROOF. Let ρ be a metric in X compatible with the topology of X. For positive integers n and m, let $E(n, m)$ denote the set of all $x \in X$ such that $\rho(x, xt) \geqq 1/m$ for all $t \in P_n$. It is clear that $X - R = \bigcup_{n,m=1}^{+\infty} E(n, m)$. For fixed positive integers n and m, the set $E(n, m)$ is closed in X. Thus $X - R$ is an F_σ set and R is a G_δ set.

Now assume that T is regionally recursive. Let n and m be fixed positive integers. Suppose int $E(n, m) \neq \emptyset$. Then there exists a nonvacuous open subset U of X such that $U \subset E(n, m)$ and $\rho(x, y) < 1/m$ $(x, y \in U)$. Since $U \cap Ut \neq \emptyset$ for some $t \in P_n$, we can find $x \in U$ such that $xt \in U$ whence $\rho(x, xt) < 1/m$. This contradicts the definition of $E(n, m)$ and therefore int $E(n, m) = \emptyset$. Thus $X - R$ is a first category subset of X and the proof is completed.

3.32. REMARK. Let X and Y be topological spaces, let A be a dense subset of X, let α be a closed subset of $Y \times Y$, and let φ and ψ be continuous maps of X into Y such that $(x\varphi, x\psi) \in \alpha$ for all $x \in A$. Then $(x\varphi, x\psi) \in \alpha$ for all $x \in X$.

3.33. THEOREM. *Let X be a uniform space, let Y be a T-invariant subset of X and let T be {recursive} {weakly$_2$ recursive} on Y. Then T is {recursive} {weakly$_2$ recursive} on $\overline{Y}$.*

PROOF. First reading. Let α be a closed index of X. There exists an admissible subset A of T such that $x \in Y$ and $t \in A$ implies $(x, xt) \in \alpha$. By 3.32, $x \in \overline{Y}$ and $t \in A$ implies $(x, xt) \in \alpha$. The proof of the first reading is completed.

Second reading. Let α be a closed index of X. There exists an admissible subset A of T and a finite subset K of T such that $y \in Y$ implies the existence of $k, h \in K$ such that $ykAh \subset y\alpha$. We show that $x \in \overline{Y}$ implies the existence of $k, h \in K$ such that $xkAh \subset x\alpha$. Let $x \in \overline{Y}$. Choose $k, h \in K$ such that for every neighborhood U of x there exists $y \in U \cap Y$ such that $ykAh \subset y\alpha$, that is, $t \in A$ implies $(y, ykth) \in \alpha$. By 3.32, $t \in A$ implies $(x, xkth) \in \alpha$. Hence, $xkAh \subset x\alpha$. The proof is completed.

3.34. THEOREM. *Let X be a uniform space, let Y be a T-invariant subset of X, let T be weakly$_1$ recursive on Y and suppose that $A, B, K \subset T$ with A admissible, K compact and $A \subset BK$ implies B is admissible. Then T is weakly$_1$ recursive on $\overline{Y}$.*

PROOF. Let α be an index of X. Choose a symmetric index β of X such that $\beta^3 \subset \alpha$. There exist an admissible subset A of T and a compact subset K of T such that $y \in Y$ implies the existence of a subset C of T such that $A \subset CK$ and $yC \subset y\beta$. We show $x \in \overline{Y}$ implies the existence of a subset B of T such that $A \subset BK$ and $xB \subset x\alpha$. It is enough to show that $x \in \overline{Y}$ and $a \in A$ implies the existence of $k \in K$ such that $xak^{-1} \in x\alpha$. Let $x \in \overline{Y}$ and let $a \in A$. By 1.20 (4), there exists an index γ of X such that $\gamma \subset \beta$ and $x\gamma ak^{-1} \subset xak^{-1}\beta$ $(k \in K)$. Choose $y \in x\gamma \cap Y$. There exists $k \in K$ such that $yak^{-1} \in y\beta$. It follows that $(x, xak^{-1}) = (x, y)(y, yak^{-1})(yak^{-1}, xak^{-1}) \in \gamma\beta^2 \subset \beta^3 \subset \alpha$ and $xak^{-1} \in x\alpha$. The proof is completed.

3.35. THEOREM. *Let X be a uniform space, let $x \in X$, let T be recursive at x, let T be equicontinuous at x, and let T be abelian. Then T is recursive on $\overline{xT}$.*

PROOF. Let α be a closed index of X. Choose an index β of X such that $x\beta\tau \subset x\tau\alpha$ $(\tau \in T)$. There exists an admissible subset A of T such that $xt \in x\beta$ $(t \in A)$. Then $x\tau t = xt\tau \in x\beta\tau \subset x\tau\alpha$ $(\tau \in T, t \in A)$ and $(x\tau, x\tau t) \in \alpha$ $(\tau \in T, t \in A)$. By 3.32, $(y, yt) \in \alpha$ $(y \in \overline{xT}, t \in A)$. The proof is completed.

3.36. INHERITANCE THEOREM. *For each subgroup G of T let there be distinguished in G certain subsets, called G-admissible, such that:*

(I) *If $A, B, C \subset T$ such that $A \subset BC$, if A is T-admissible, and if C is compact, then B is T-admissible.*

(II) *If G is a closed syndetic subgroup of T and if $A \subset G$, then A is G-admissible if and only if A is T-admissible.*

Let T be locally compact and let S be a closed syndetic invariant subgroup of T. Then:

(1) *If $x \in X$, then S is recursive at x if and only if T is recursive at x.*

(2) *S is pointwise recursive if and only if T is pointwise recursive.*

PROOF. We first show that if T is recursive at x, then S_x is recursive at x. Suppose T is recursive at x. Let U be a neighborhood of x. There exist neighborhoods V of x and W of e such that $W = W^{-1}$, W is compact and $VW \subset U$. By 2.10(2), we may suppose that $V \cap x(T - S_xW) = \emptyset$. There exists a T-admissible subset A of T such that $xA \subset V$. Clearly $A \subset S_xW$ and $xAW \subset U$. Define $B = S_x \cap AW$. Since $A \subset BW$, B is a T-admissible subset of T. Also $B \subset S_x$ and $xB \subset U$. Thus S_x is recursive at x.

We next show that if S_x is recursive at x, then S is recursive at x. Suppose S_x is recursive at x. Let U be an open neighborhood of x. By 2.10(3), there exists a compact subset M of T such that $xM \subset U$ and $S_x \subset SM^{-1}$. Let V be a neighborhood of x for which $VM \subset U$. There exists an S_x-admissible subset A of S_x such that $xA \subset V$. Now $xAM \subset U$. Define $B = S \cap AM$. Since $A \subset BM^{-1}$, B is a T-admissible subset of T. Also $B \subset S$ and $xB \subset U$. Thus S is recursive at x.

It now follows that if T is recursive at x, then S is recursive at x. The converse is obvious. The proof is completed.

3.37. DEFINITION. Let T be a topological group. A subset S of T is said to be *replete in T*, provided that S contains some bilateral translate of each compact subset of T. A subset A of T is said to be *extensive in T* provided that A intersects every replete semigroup in T.

3.38. DEFINITION. If in 3.13 the term *admissible set* is replaced by {*left syndetic set*}{*syndetic invariant subgroup*}{*translated syndetic invariant subgroup*}{*extensive set*}, then the term *recursive* is replaced by {*almost periodic*}{*regularly almost periodic*}{*isochronous*}{*recurrent*}. By {*weakly almost periodic*}{*weakly isochronous*} we shall mean {weakly$_1$ almost periodic}{weakly$_2$ isochronous}.

3.39. Remark. The choice of "left" rather than "right" in the first reading of 3.38 is dictated by the notation adopted in 1.01 and 3.05, namely,

$$\pi : X \times T \to X,$$

$$xt = (x, t)\pi \qquad (x \in X, t \in T)$$

whence the homomorphism axiom takes the form

$$(xt)s = x(ts) \qquad (x \in X; t, s \in T).$$

If the adopted notation were

$$\pi : T \times X \to X,$$

$$tx = \pi(t, x) \qquad (t \in T, x \in X)$$

with the homomorphism axiom written as

$$s(tx) = (st)x \qquad (t, s \in T; x \in X),$$

then we would choose "right" rather than "left". In the latter notation the order of terms in the image of a point under a mapping and in a mapping product would be the reverse of the order in the former notation. An isomorphism between the two notations is established by the group inversion of T.

3.40. Remark. The recursion properties of 3.38 are intrinsic. They include all of the recursion properties that we shall study.

3.41. Remark. Let *admissible* and *recursive* refer to one of the specializations of 3.38. If $\mathfrak{I}$ and $\mathfrak{S}$ are topologies of the group T such that $\mathfrak{I} \subset \mathfrak{S}$, if $A \subset T$ and if A is admissible relative to $\mathfrak{S}$, then A is admissible relative to $\mathfrak{I}$. Hence the strongest recursion properties are those relative to the discrete topology of T.

3.42. Definition. Let φ be a function on a topological group T to a uniform space Y.

The function φ is said to be {*left*}{*right*} *uniformly recursive* provided that if α is an index of Y, then there exists an admissible subset A of T such that $\tau \in T$ and $t \in A$ implies $\{(\tau\varphi, t\tau\varphi) \in \alpha\}\{(\tau\varphi, \tau t^{-1}\varphi) \in \alpha\}$.

The function φ is said to be {*left*}{*right*} *weakly recursive* provided that if E is a compact subset of T and if α is an index of Y, then there exists an admissible subset A of T such that $\tau \in E$ and $t \in A$ implies $\{(\tau\varphi, t\tau\varphi) \in \alpha\}$ $\{(\tau\varphi, \tau t^{-1}\varphi) \in \alpha\}$.

The function φ is said to be *bilaterally* {*uniformly*}{*weakly*} *recursive* provided that φ is both left and right {*uniformly*}{*weakly*} recursive.

3.43. Remark. We adopt the notation of 1.62. Let $\varphi \in \Phi$. Then φ is a {left}{right} uniformly recursive function if and only if φ is a recursive point under (Φ, T, ρ).

We adopt the notation of 1.63. Let $\varphi \in \Phi$. Then φ is a {left}{right} weakly recursive function if and only if φ is a recursive point under (Φ, T, ρ).

3.44. Definition. Let Φ be a class of functions on a topological group T to a uniform space Y.

The function class Φ is said to be *{left} {right} uniformly recursive* provided that if α is an index of Y, then there exists an admissible subset A of T such that $\tau \in T$, $t \in A$ and $\varphi \in \Phi$ implies $\{(\tau\varphi, t\tau\varphi) \in \alpha\}\{(\tau\varphi, \tau t^{-1}\varphi) \in \alpha\}$.

The function class Φ is said to be *{left} {right} weakly recursive* provided that if E is a compact subset of T and if α is an index of Y, then there exists an admissible subset A of T such that $\tau \in E$, $t \in A$ and $\varphi \in \Phi$ implies $\{(\tau\varphi, t\tau\varphi) \in \alpha\}\{(\tau\varphi, \tau t^{-1}\varphi) \in \alpha\}$.

The function class Φ is said to be *bilaterally {uniformly} {weakly} recursive* provided that Φ is both left and right {uniformly} {weakly} recursive.

3.45. Remark. We adopt the notation of 1.62. Let $\Psi \subset \Phi$ and let Ψ be invariant under (Φ, T, ρ). Then the function class Ψ is {left} {right} uniformly recursive if and only if (Φ, T, ρ) is recursive on the set Ψ.

We adopt the notation of 1.63. Let $\Psi \subset \Phi$ and let Ψ be invariant under (Φ, T, ρ). Then the function class Ψ is {left} {right} weakly recursive if and only if (Φ, T, ρ) is recursive on the set Ψ.

3.46. Remark. In the definition of right uniformly recursive in 3.42 and 3.44, the expression $(\tau\varphi, \tau t^{-1}\varphi) \in \alpha$ may be replaced by the expression $(\tau\varphi, \tau t\varphi) \in \alpha$ without changing the sense of the definition.

3.47. Definition. Let φ be a function on X to a uniform space Y.

The function φ is said to be *uniformly recursive relative to* (X, T, π) provided that if α is an index of Y, then there exists an admissible subset A of T such that $x \in X$ and $t \in A$ implies $(x\varphi, x\pi^{t^{-1}}\varphi) \in \alpha$.

The function φ is said to be *weakly recursive relative to* (X, T, π) provided that if E is a compact subset of X and if α is an index of Y, then there exists an admissible subset A of T such that $x \in E$ and $t \in A$ implies $(x\varphi, x\pi^{t^{-1}}\varphi) \in \alpha$.

3.48. Remark. Let φ be a function on a topological group T to a uniform space Y. Then:

(1) φ is uniformly recursive relative to the {left} {right} transformation group of T if and only if φ is {left} {right} uniformly recursive.

(2) φ is weakly recursive relative to the {left} {right} transformation group of T if and only if φ is {left} {right} weakly recursive.

3.49. Remark. We adopt the notation of {1.66} {1.68}. Let $\varphi \in \Phi$. Then φ is a {uniformly} {weakly} recursive function relative to (X, T, π) if and only if φ is a recursive point under (Φ, T, ρ).

3.50. Definition. Let Φ be a class of functions on X to a uniform space Y.

The function class Φ is said to be *uniformly recursive relative to* (X, T, π) provided that if α is an index of Y, then there exists an admissible subset A of T such that $x \in X$, $t \in A$ and $\varphi \in \Phi$ implies $(x\varphi, x\pi^{t^{-1}}\varphi) \in \alpha$.

The function class Φ is said to be *weakly recursive relative to* (X, T, π) provided that if E is a compact subset of X and if α is an index of Y, then there exists an admissible subset A of T such that $x \in E$, $t \in A$ and $\varphi \in \Phi$ implies $(x\varphi, x\pi^{t^{-1}}\varphi) \in \alpha$.

3.51. REMARK. Let Φ be a class of functions on a topological group T to a uniform space Y. Then:

(1) Φ is uniformly recursive relative to the {left}{right} transformation group of T if and only if Φ is {left}{right} uniformly recursive.

(2) Φ is weakly recursive relative to the {left}{right} transformation group of T if and only if Φ is {left}{right} weakly recursive.

3.52. REMARK. We adopt the notation of {1.66}{1.68}. Let $\Psi \subset \Phi$ and let Ψ be invariant under (Φ, T, ρ). Then the function class Ψ is {uniformly} {weakly} recursive relative to (X, T, π) if and only if (Φ, T, ρ) is recursive on the set Ψ.

3.53. REMARK. Let X be a uniform space. Then:

(1) If $x \in X$, then the transformation group (X, T, π) is recursive at x if and only if the motion $\pi_x : T \to X$ is left weakly recursive.

(2) If $x \in X$, then the transformation group (X, T, π) is recursive on xT if and only if the motion $\pi_x : T \to X$ is right uniformly recursive.

(3) The transformation group (X, T, π) is recursive if and only if the motion space $[\pi_x : T \to X \mid x \in X]$ is right uniformly recursive.

3.54. REMARK. Let X be a uniform space and let φ be a function on X to a uniform space Y. Then:

(1) If $x \in X$, if the transformation group (X, T, π) is recursive at x and if φ is continuous on xT, then the function $\pi_x\varphi : T \to Y$ is left weakly recursive.

(2) If $x \in X$, if the transformation group (X, T, π) is recursive on xT and if φ is uniformly continuous on xT, then the function $\pi_x\varphi : T \to Y$ is right uniformly recursive.

(3) If the transformation group (X, T, π) is recursive and if φ is uniformly continuous, then the function class $[\pi_x\varphi : T \to Y \mid x \in X]$ is right uniformly recursive.

3.55. NOTES AND REFERENCES.

(3.13) Use of the term *recursive* in studying simultaneously a number of the diverse recurrence phenomena which are of interest in the analysis of transformation groups occurs in Gottschalk and Hedlund [5]. The expression *weakly almost periodic* was introduced by Gottschalk [6].

(3.36) Cf. Gottschalk [2, 6, 8], Erdös and Stone [1], Gottschalk and Hedlund [5].

(3.38) The terms *replete* and *extensive*, as defined here, were introduced by Gottschalk and Hedlund [10]. If T is either $\mathcal{I}$ or $\mathcal{R}$, a subset A of T is extensive if and only if A contains a sequence marching to $+\infty$ and a sequence marching to $-\infty$.

The expression *almost periodic*, as applied to a point, is a generalization of the term *recurrent* as used by G. D. Birkhoff ([1], vol. 1, pp. 654–672) for the case of a continuous flow in a compact space. It is not unrelated to the classic terminology of Bohr. A function which is almost periodic in the sense of Bohr is an almost periodic point of the continuous flow defined by translation if the function space under consideration is provided with the uniform (space-index) topology, and conversely. The expression *almost periodic*, as applied to a transformation group, is closely related to the same expression as used by Cameron [1] and Montgomery [1] (see, in this connection, 4.38). The phrase *regularly almost periodic* was introduced by G. T. Whyburn [1] in connection with a continuous transformation and its iterates. As applied to a transformation group, it is closely related to the *nearly periodic* of P. A. Smith [1]. In the case of a discrete or continuous flow, the property of *recurrence* has had a varied nomenclature (cf. Hedlund [4]). The term *isochronous* is to be found in García and Hedlund [1].

4. ALMOST PERIODICITY

4.01. STANDING NOTATION. Throughout this section (X, T, π) denotes a transformation group.

4.02. REMARK. Let $x \in X$. Then the following statements are pairwise equivalent:

(1) T is almost periodic at x; that is to say, if U is a neighborhood of x, then there exists a left syndetic subset A of T such that $xA \subset U$.

(2) If U is a neighborhood of x, then there exists a compact subset K of T such that $t \in T$ implies $xtK \cap U \neq \emptyset$.

(3) If U is a neighborhood of x, then there exists a compact subset K of T such that $xT \subset UK$.

4.03. THEOREM. *Let $x \in X$ and let T be almost periodic at x. Then T is almost periodic at every point of xT.*

PROOF. Use 3.21.

4.04. INHERITANCE THEOREM. *Let T be locally compact and let S be a closed syndetic invariant subgroup of T. Then:*

(1) *If $x \in X$, then S is almost periodic at x if and only if T is almost periodic at x.*

(2) *S is pointwise almost periodic if and only if T is pointwise almost periodic.*

PROOF. Use 3.36.

4.05. THEOREM. *Let M be a compact minimal orbit-closure under T. Then T is discretely almost periodic at every point of M.*

PROOF. Let $x \in M$ and let U be an open neighborhood of x. By 2.12, $M \subset UT$. There exists a finite subset K of T such that $M \subset UK$ whence $xT \subset UK$. By 4.02, x is discretely almost periodic.

4.06. THEOREM. *Let X be compact. Then there exists a point of X which is discretely almost periodic under T.*

PROOF. Use 2.22 and 4.05.

4.07. THEOREM. *Let X be regular, let $x \in X$ and let T be almost periodic at x. Then $\overline{xT}$ is minimal under T.*

PROOF. Assume $\overline{xT}$ is not minimal. Then there exists $y \in \overline{xT}$ such that $x \notin \overline{yT}$. Let U be a closed neighborhood of x for which $U \cap yT = \emptyset$ and let K be a compact subset of T for which $xT \subset UK$. Since $yK^{-1} \cap U = \emptyset$, there exists by 1.18 (4) a neighborhood V of y such that $VK^{-1} \cap U = \emptyset$ whence $V \cap UK = \emptyset$. However, $y \in \overline{xT}$, so that $xT \cap V \neq \emptyset$. Since also $xT \subset UK$, we have $V \cap UK \neq \emptyset$. This is a contradiction. The proof is completed.

4.08. THEOREM. *Let X be regular and let T be pointwise almost periodic. Then the class of all orbit-closures under T is a partition of X.*

PROOF. Use 4.07 and 2.23.

4.09. THEOREM. *Let X be a T_2-space, let $x \in X$, let there exist a compact neighborhood of x and let T be almost periodic at x. Then $\overline{xT}$ is compact and T is discretely almost periodic at every point of $\overline{xT}$.*

PROOF. Let U be a compact neighborhood of x and let K be a compact subset of T such that $\overline{xT} \subset UK$. Since UK is compact, the conclusion follows from 4.05 and 4.07.

4.10. THEOREM. *Let X be a locally compact T_2-space. Then the following statements are pairwise equivalent:*

(1) *T is pointwise almost periodic.*

(2) *T is discretely pointwise almost periodic.*

(3) *The class of all orbit-closures under T is a decomposition of X.*

PROOF. Use 4.05, 4.08 and 4.09.

4.11. THEOREM. *Let X be regular, let $x \in X$ and let T be locally almost periodic at x. Then T is locally almost periodic at every point of $\overline{xT}$.*

PROOF. Use 3.25 and 4.07.

4.12. REMARK. Let $x \in X$. Then the following statements are pairwise equivalent:

(1) T is locally weakly almost periodic at x; that is to say, if U is a neighborhood of x, then there exist a neighborhood V of x, a left syndetic subset A of T and a compact subset C of T such that $y \in V$ implies the existence of a subset B of T for which $A \subset BC$ and $yB \subset U$.

(2) If U is a neighborhood of x, then there exist a neighborhood V of x and a compact subset K of T such that $y \in V$ implies the existence of a subset A of T such that $T = AK$ and $yA \subset U$.

(3) If U is a neighborhood of x, then there exist a neighborhood V of x and a compact subset K of T such that $y \in V$ and $t \in T$ implies $ytK \cap U \neq \emptyset$.

(4) If U is a neighborhood of x, then there exist a neighborhood V of x and a compact subset K of T such that $VT \subset UK$.

4.13. THEOREM. *Let $x \in X$ and let T be locally weakly almost periodic at x. Then T is locally weakly almost periodic at every point of xT.*

PROOF. Use 3.21.

4.14. THEOREM. *Let the class of all orbit-closures under T be a star-closed decomposition of X. Then T is discretely locally weakly almost periodic.*

PROOF. Let $x \in X$ and let U be an open neighborhood of x. Since $\overline{xT} \subset UT$ by 2.12 and $\overline{xT}$ is compact, there exists a finite subset K of T such that $\overline{xT} \subset UK$.

By 1.36 there exists a neighborhood V of x for which $VT \subset UK$. The conclusion now follows from 4.12.

4.15. THEOREM. *Let X be a T_2-space, let E be a compact subset of X, let there exist a compact neighborhood of each point of E and let T be locally weakly almost periodic at each point of E. Then $\overline{ET}$ is compact.*

PROOF. For each $x \in E$ choose a compact neighborhood U_x of x, a neighborhood V_x of x and a compact subset K_x of T such that $V_xT \subset U_xK_x$. There exists a finite subset F of E such that $E \subset \bigcup_{x\in F} V_x$. Since $ET \subset \bigcup_{x\in F} V_xT \subset \bigcup_{x\in F} U_xK_x$, the proof is completed.

4.16. THEOREM. *Let X be a locally compact T_2-space and let T be locally weakly almost periodic. Then the class of all orbit-closures under T is a star-closed decomposition of X.*

PROOF. By 4.15 we may assume that X is compact. Let A be an orbit-closure under T and let U be an open neighborhood of A. It is enough to show that there exists a neighborhood V of A such that $VT \subset U$. Choose a closed neighborhood W of A such that $W \subset U$. For each $x \in X - U$ there exists a neighborhood N_x of x and a compact subset K_x of T such that $N_xT \subset (X - W)K_x$. Select a finite subset E of $X - U$ so that $X - U \subset \bigcup_{x\in E} N_x$. Define $K = \bigcup_{x\in E} K_x$. It follows that K is a compact subset of T for which $(X - U)T \subset (X - W)K$. Choose a neighborhood V of A such that $VK^{-1} \subset W$. Then $VK^{-1} \cap (X - W) = \emptyset$, $V \cap (X - W)K = \emptyset$, $V \cap (X - U)T = \emptyset$, $VT \cap (X - U) = \emptyset$ and $VT \subset U$. The proof is completed.

4.17. THEOREM. *Let X be a locally compact T_2-space. Then the following statements are pairwise equivalent:*

(1) *T is locally weakly almost periodic.*

(2) *T is discretely locally weakly almost periodic.*

(3) *The class of all orbit-closures under T is a star-closed decomposition of X.*

PROOF. Use 4.14 and 4.16.

4.18. THEOREM. *Let X be a T_2-space, let $x \in X$, let there exist a compact neighborhood of x and let T be almost periodic at x. Then T is discretely locally weakly almost periodic on $\overline{xT}$.*

PROOF. Use 4.07, 4.09 and 4.14.

4.19. **INHERITANCE THEOREM.** *Let X be a compact T_2-space, let X be minimal under T and let S be a syndetic invariant subgroup of T. Then S is locally weakly almost periodic.*

PROOF. Use 2.32 and 4.17.

4.20. STANDING NOTATION. For the remainder of this section X denotes a separated uniform space.

4.21. REMARK. The following statements are pairwise equivalent:

(1) T is weakly almost periodic; that is to say, if α is an index of X, then there exist a left syndetic subset A of T and a compact subset C of T such that $x \in X$ implies the existence of a subset B of T for which $A \subset BC$ and $xB \subset x\alpha$.

(2) If α is an index of X, then there exists a compact subset K of T such that $x \in X$ implies the existence of a subset A of T for which $T = AK$ and $xA \subset x\alpha$.

(3–4–5–6) If α is an index of X, then there exists a compact subset K of T such that $x \in X$ and $t \in T$ implies $\{xtK \cap x\alpha \neq \emptyset\}\{xK \cap xt\alpha \neq \emptyset\}\{xt \in x\alpha K\}\{xt \in xK\alpha\}$.

(7–8) If α is an index of X, then there exists a compact subset K of T such that $x \in X$ implies $\{xT \subset x\alpha K\}\{xT \subset xK\alpha\}$.

(9–10) If α is an index of X, then there exist an index β of X and a compact subset K of T such that $x \in X$ implies $\{x\beta T \subset x\alpha K\}\{xT\beta \subset xK\alpha\}$.

4.22. REMARK. Let T be weakly almost periodic. Then every orbit under T, and therefore every orbit-closure under T, is totally bounded.

PROOF. Use 4.21 (8).

4.23. REMARK. The following statements are valid:

(1) If T is weakly almost periodic, then T is locally weakly almost periodic.

(2) If X is compact and if T is locally weakly almost periodic, then T is weakly almost periodic.

4.24. THEOREM. *Let X be compact. Then the following statements are pairwise equivalent:*

(1) *T is weakly almost periodic.*

(2) *T is discretely weakly almost periodic.*

(3) *The class of all orbit-closures under T is a star-closed decomposition of X.*

PROOF. Use 4.17 and 4.23.

4.25. REMARK. Consider the following statements:

(I–II) If α is an index of X, then there exist an index β of X and a compact subset K of T such that $x \in X$ implies $\{x\beta T \subset xK\alpha\}\{xT\beta \subset x\alpha K\}$.

(III–IV) If α is an index of X, then there exists an index β of X such that $x \in X$ implies $\{x\beta T \subset xT\alpha\}\{xT\beta \subset x\alpha T\}$.

(V) T is weakly almost periodic.

Then:

(1) I is equivalent to II.

(2) III is equivalent to IV.

(3) I is equivalent to the conjunction of III and V.

(4) If X is compact, then I through V are pairwise equivalent.

4.26. THEOREM. *Let $x \in X$, let there exist a compact neighborhood of x and let T be almost periodic at x. Then T is discretely weakly almost periodic on $\overline{xT}$.*

Proof. Use 4.09 and 4.24.

4.27. **Theorem.** *Let Y be a T-invariant subset of X and let T be weakly almost periodic on Y. Then T is weakly almost periodic on $\overline{Y}$.*

Proof. Use 3.34.

4.28. **Theorem.** *Let X be complete, let $x \in X$ and let T be weakly almost periodic on xT. Then $\overline{xT}$ is compact and T is discretely weakly almost periodic on $\overline{xT}$.*

Proof. Use 4.22, 4.24 and 4.27.

4.29. **Theorem.** *Let X be compact, let X be minimal under T and let S be a syndetic invariant subgroup of T. Then S is weakly almost periodic.*

Proof. Use 2.32 and 4.24.

4.30. **Theorem.** *Let X be a compact minimal orbit-closure under T and let $\mathcal{A}$ be a class of nonvacuous subsets of T, called admissible, such that $t\mathcal{A}s \subset \mathcal{A}$ $(t, s \in T)$. Consider the following statements:*

(I) T is locally recursive.

(II) T is locally recursive at some point of X.

(III) If α is an index of X, then there exist $x \in X$, a neighborhood V of x and an admissible subset A of T such that $VA \subset x\alpha$.

(IV) T is weakly$_2$ recursive.

Then:

(1) I *is equivalent to* II; II *implies* III; III *is equivalent to* IV.

(2) *If X is metrizable, then* I, II, III, IV *are pairwise equivalent.*

Proof. (1) By 3.25, I is equivalent to II. Clearly, II implies III.

Assume III. We prove IV. Let α be an index of X. Choose a symmetric index β of X for which $\beta^3 \subset \alpha$. By 4.21 (7) and 4.24 there exists a finite subset F of T such that $x \in X$ implies $xT \subset x\beta F^{-1}$. Select an index γ of X such that $x \in X$ and $s \in F$ implies $x\gamma s \subset xs\beta$. There exist $x_0 \in X$, a neighborhood V of x_0 and an admissible subset A of T such that $VA \subset x_0\gamma$. Choose a finite subset E of T for which $X = VE^{-1}$. Define $K = E \cup F$. We show $x \in X$ implies the existence of $k, h \in K$ such that $xkAh \subset x\alpha$. Let $x \in X$. Choose $k \in E$ such that $xk \in V$ and then choose $h \in F$ such that $xkAh \cap x\beta \neq \emptyset$. Since $xkAh \subset VAh \subset x_0\gamma h \subset x_0h\beta$, it follows that $xkAh \subset x\beta^3 \subset x\alpha$. This proves IV.

Assume IV. We prove III. Let α be an index of X. Choose a symmetric index β of X for which $\beta^4 \subset \alpha$. There exist an admissible subset B of T and a finite subset K of T such that $x \in X$ implies the existence of $k, h \in K$ such that $xkBh \subset x\beta$. For $k, h \in K$ let $\{k, h\}$ denote the set of all $x \in X$ such that $xkBh \subset x\beta$. Clearly $X = \bigcup_{k,h\in K} \overline{\{k, h\}}$.

We show that if $k, h \in K$ and if $x \in \overline{\{k, h\}}$, then $xkBh \subset x\beta^3$. Let $k, h \in K$, let $x \in \overline{\{k, h\}}$ and let $b \in B$. Choose an index γ of X such that $\gamma \subset \beta$ and $x\gamma kbh \subset xkbh\beta$. Now choose $y \in x\gamma \cap \{k, h\}$. Then $(x, xkbh) = (x, y)(y, ykbh)(ykbh, xkbh) \in \gamma\beta^2 \subset \beta^3$ and $xkbh \in x\beta^3$.

Since $X = \bigcup_{k,h \in K} \overline{\{k, h\}}$ and K is finite, there exist $k, h \in K$ such that int $\overline{\{k, h\}} \neq \emptyset$. Hence there exist $k, h \in K$, $x \in \overline{\{k, h\}}$ and a neighborhood V of x such that $V \times V \subset \beta$ and $V \subset \overline{\{k, h\}}$. Now $VkBh \subset x\alpha$ since $y \in V$ and $b \in B$ implies $(x, ykbh) = (x, y)(y, ykbh) \in \beta^4 \subset \alpha$ whence $ykbh \in x\alpha$. Define $A = kBh$. Then A is an admissible subset of T such that $VA \subset x\alpha$. This proves III.

(2) Assume III. We prove II.

We first show: (L) If U is a nonvacuous open subset of X, then there exist $x \in U$, a neighborhood V of x and an admissible subset A of T such that $VA \subset U$.

Let U be a nonvacuous open subset of X. Since $X = UT$, there exists an index α of X such that $x \in X$ implies the existence of $t \in T$ such that $x\alpha \subset Ut$ whence $x\alpha t^{-1} \subset U$. There exist $y \in X$, a neighborhood W of y and an admissible subset B of T such that $WB \subset y\alpha$. Choose $t \in T$ such that $y\alpha t^{-1} \subset U$. Define $x = yt^{-1}$. Since $xt = y$, there exists a neighborhood V of x such that $Vt \subset W$. Hence $VtBt^{-1} \subset WBt^{-1} \subset y\alpha t^{-1} \subset U$. Define $A = tBt^{-1}$. Since A is an admissible subset of T such that $VA \subset U$, the proof of (L) is completed.

Let $[\alpha_n \mid n = 1, 2, \cdots]$ be a countable base of the uniformity of X. Define $U_0 = X$. For $n = 1, 2, \cdots$, we proceed inductively as follows:

By (L) there exists a nonvacuous open subset U_n of U_{n-1} and an admissible subset A_n of T such that $U_n \times U_n \subset \alpha_n$, $\overline{U}_n \subset U_{n-1}$ and $U_n A_n \subset U_{n-1}$.

It is clear that T is locally recursive at every point of $\bigcap_{n=1}^{+\infty} U_n$. Since $\bigcap_{n=1}^{+\infty} U_n \neq \emptyset$, the proof is completed.

4.31. THEOREM. *Let X be a compact minimal orbit-closure under T. Consider the following statements*:

(I) *T is locally almost periodic.*

(II) *T is locally almost periodic at some point of X.*

(III) *If α is an index of X, then there exist $x \in X$, a neighborhood V of x and a left syndetic subset A of T such that $VA \subset x\alpha$.*

(IV) *T is weakly$_2$ almost periodic.*

Then:

(1) I *is equivalent to* II; II *implies* III; III *is equivalent to* IV.

(2) *If X is metrizable, then* I, II, III, IV *are pairwise equivalent.*

PROOF. Use 4.30.

4.32. REMARK. The following statements are pairwise equivalent:

(1) T is almost periodic; that is to say, if α is an index of X, then there exists a left syndetic subset A of T such that $x \in X$ implies $xA \subset x\alpha$.

(2–3–4–5) If α is an index of X, then there exists a compact subset K of T such that to each $t \in T$ there corresponds $k \in K$ such that $x \in X$ implies $\{xtk \in x\alpha\}\{xk \in xt\alpha\}\{xt \in x\alpha k\}\{xt \in xk\alpha\}$.

4.33. REMARK. The transformation group T is discretely almost periodic if and only if the transition group $[\pi^t \mid t \in T]$ is totally bounded in its space-index uniformity.

4.34. THEOREM. *Let Y be a T-invariant subset of X and let T be almost periodic on Y. Then T is almost periodic on $\overline{Y}$.*

PROOF. Use 3.33.

4.35. THEOREM. *Let X be compact and let T be almost periodic. Then T is discretely almost periodic.*

PROOF. Let α be an index of X. Choose an index β of X so that $\beta^2 \subset \alpha$. There exists a compact subset K of T such that $t \in T$ implies $(xk, xt) \in \beta$ $(x \in X)$ for some $k \in K$. Since $\pi : X \times K \to X$ is uniformly continuous, $[\pi^k \mid k \in K]$ is equicontinuous. By 11.12, $[\pi^k \mid k \in K]$ is totally bounded in its space-index uniformity. Hence, there exists a finite subset F of K such that $k \in K$ implies $(xf, xk) \in \beta$ $(x \in X)$ for some $f \in F$. If $t \in T$, then there exist $k \in K$, $f \in F$ such that $(xk, xt) \in \beta$ $(x \in X)$, $(xf, xk) \in \beta$ $(x \in X)$ whence $(xf, xt) \in \alpha$ $(x \in X)$. The proof is completed.

4.36. DEFINITION. The transformation group T is said to be *uniformly continuous* provided that every transition π^t $(t \in T)$ is uniformly continuous.

4.37. THEOREM. *Let T be uniformly continuous and discretely almost periodic. Then T is uniformly equicontinuous.*

PROOF. Use 4.33 and 11.12.

4.38. THEOREM. *Let X be totally bounded and let T be uniformly continuous. Then the following statements are pairwise equivalent:*

(1) *T is discretely almost periodic.*

(2) *T is uniformly equicontinuous.*

(3) *The transition group $[\pi^t \mid t \in T]$ is totally bounded in its space-index uniformity.*

(4) *The motion space $[\pi_x \mid x \in X]$ is totally bounded in its space-index uniformity.*

(5) *If α is an index of X, then there exists a finite partition $\mathfrak{A}$ of X such that $A \in \mathfrak{A}$ and $t \in T$ implies $At \times At \subset \alpha$.*

(6) *If α is an index of X, then there exists a finite partition $\mathfrak{B}$ of T such that $x \in X$ and $B \in \mathfrak{B}$ implies $xB \times xB \subset \alpha$.*

PROOF. By 4.37, (1) implies (2).

By 11.12, (2) is equivalent to (3).

We show (3) implies (1). Assume (3). Let α be an index of X. There exists a finite subset F of T such that $t \in T$ implies $(xf, xt) \in \alpha$ $(x \in X)$ for some $f \in F$. This proves (1).

By 11.06, (3) is equivalent to (5), and (4) is equivalent to (6). By 11.23 (1), (2) implies (4). By 11.23 (2), (4) implies (5). The proof is completed.

4.39. REMARK. The following statements are valid:

(1) If X is compact and if α is an index of X, then there exist an index β of X and a neighborhood V of e such that $x \in X$ implies $x\beta V \times x\beta V \subset \alpha$.

(2) If X is compact and if T is provided with its left uniformity, then the motion space $[\pi_x \mid x \in X]$ is uniformly equicontinuous.

4.40. **Inheritance Theorem.** *Let S be a subgroup of T. If T is discretely almost periodic or if X is compact and T is almost periodic, then S is discretely almost periodic.*

Proof. Use 4.35 and 4.33.

4.41. Lemma. *Let X be a uniform space and let $A \subset X$. Then A is totally bounded if and only if for each index α of X there exists a totally bounded subset E of X such that $A \subset E\alpha$.*

Proof. The necessity is obvious. We prove the sufficiency. Let α be an index of X. Choose an index β of X such that $\beta^2 \subset \alpha$. There exists a totally bounded subset E of X such that $A \subset E\beta$. Select a finite subset F of E such that $E \subset F\beta$. Then $A \subset E\beta \subset F\beta^2 \subset F\alpha$. The proof is completed.

4.42. Theorem. *Let $x \in X$, let x be almost periodic under T and let T be equicontinuous at x. Then:*

(1) *xT is totally bounded.*

(2) *If T is abelian, then T is almost periodic on $\overline{xT}$.*

Proof. (1) We use 4.41. Let α be an index of X. There exists an index β of X such that $x\beta t \subset xt\alpha$ $(t \in T)$. For some compact subset K of T, we have $xT \subset x\beta K$. It follows that $xT \subset x\beta K \subset xK\alpha$.

(2) Use 3.35.

4.43. Theorem. *Let $x \in X$ and let T be uniformly equicontinuous. Then the following statements are pairwise equivalent:*

(1) *x is almost periodic under T.*

(2) *xT is totally bounded.*

(3) *T is discretely almost periodic on xT.*

Proof. Use 4.38 and 4.42.

4.44. Theorem. *Let X be complete, let $x \in X$ and let T be uniformly equicontinuous on xT. Then x is almost periodic under T if and only if $\overline{xT}$ is compact.*

Proof. Use 4.43.

4.45. Theorem. *Let X be compact, let Φ be the total homeomorphism group of X and let Φ be provided with its space-index topology. Then the following statements are equivalent:*

(1) *(X, T, π) is almost periodic.*

(2) *The closure in Φ of the transition group $[\pi^t : X \to X \mid t \in T]$ is a compact topological group.*

Proof. Use 4.35, 4.38 and 11.19.

4.46. Remark. Let φ be a continuous homomorphism of a topological group T into a compact topological group S and let (S, T, ρ) be the {left}{right} transformation group of S induced by T under φ.

Then:

(1) (S, T, ρ) is almost periodic.

(2) S is an almost periodic minimal orbit-closure under (S, T, ρ) if and only if $\overline{T\varphi} = S$.

4.47. DEFINITION. Let G be a topological group. A *compactification* of G is a couple (H, φ) consisting of a compact group H and a continuous homomorphism $\varphi : G \to H$ such that $\overline{G\varphi} = H$.

4.48. THEOREM. *Let $x \in X$ and let T be abelian. Then the following statements are pairwise equivalent:*

(1) *X is an almost periodic compact minimal orbit-closure under (X, T, π).*

(2) *There exists a unique group structure of X which makes X a topological group such that (X, π_x) is a compactification of T.*

(3–4) *There exist a compact topological group S and a continuous homomorphism $\varphi : T \to S$ such that $\overline{T\varphi} = S$ and the {left}{right} transformation group of S induced by T under φ is isomorphic to (X, T, π).*

PROOF. Assume (1). We prove (2). Let Φ be the total homeomorphism group of X, let Φ be provided with its space-index uniformity and let $\Phi_0 = [\pi^t \mid t \in T]$. By 4.45, $\Psi = \overline{\Phi_0}$ is a compact abelian topological group. Define the continuous mapping

$$f : \Psi \overset{\text{onto}}{\to} X$$

by $\varphi f = x\varphi$ $(\varphi \in \Psi)$. Let P be the period of x under Ψ, that is, the greatest subset P of Ψ such that $xP = x$. Now P is a subgroup of Ψ and $\Psi/P = [yf^{-1} \mid y \in X]$ is a topological group. Define the homeomorphism

$$h : X \overset{\text{onto}}{\to} \Psi/P$$

by $yh = yf^{-1}$ $(y \in X)$. The conclusion follows.

To complete the proof, use 4.46.

4.49. DEFINITION. A topological group G is said to be {*monothetic*}{*solenoidal*} provided there exists a continuous homomorphism $\{\varphi : \mathcal{I} \to G\}$ $\{\varphi : \mathcal{R} \to G\}$ such that $\{\overline{\mathcal{I}\varphi} = G\}\{\overline{\mathcal{R}\varphi} = G\}$.

4.50. REMARK. Let G be a T_2 topological group. Then:

(1) If G is monothetic or solenoidal, then G is abelian.

(2) G is monothetic if and only if there exists $x \in G$ such that $[x^n \mid n \in \mathcal{I}]$ is dense in G.

4.51. REMARK. Let X be a compact uniform space. Then the following statements are equivalent:

(1) There exists a {discrete}{continuous} flow on X under which X is an almost periodic minimal orbit-closure.

(2) There exists a group structure of X which makes X a {monothetic} {solenoidal} topological group.

4.52. THEOREM. *Let G be a compact abelian group. Then the following statements are pairwise equivalent:*

(1) *G is monothetic.*

(2) *G is separable and G/K is monothetic where K is the identity component of G.*

(3) *G is separable and if H is an open-closed finite-indexed subgroup of G, then G/H is cyclic.*

(4) *The character group of G is algebraically isomorphic to a subgroup of the circle group.*

PROOF. Cf. Anzai and Kakutani [2].

4.53. REMARK. Let G be a compact connected separable abelian group. Then G is monothetic.

4.54. THEOREM. *Let G be a compact abelian group. Then the following statements are pairwise equivalent:*

(1) *G is solenoidal.*

(2) *G is separable connected.*

(3) *The character group of G is algebraically isomorphic to a subgroup of the line group $\mathcal{R}$.*

PROOF. Cf. Anzai and Kakutani [2].

4.55. THEOREM. *Let X be a compact uniform space containing more than one point and let X be an almost periodic minimal orbit-closure under a continuous flow $(X, \mathcal{R}, \pi)$. Then X is not totally minimal under $(X, \mathcal{R}, \pi)$.*

PROOF. Let $x \in X$. By **4.48** we may suppose that X is a topological group such that $\pi_x : \mathcal{R} \to X$ is continuous homomorphic. By the theory of characters there exists a continuous homomorphism φ of X onto the (additive) circle group C. There exists $t \in \mathcal{R}$ such that the subgroup of C generated by $xt\varphi$ is finite. Since $[x(tn) \mid n \in \mathcal{J}]\varphi = [(tn)\pi_x\varphi \mid n \in \mathcal{J}] = [(t\pi_x\varphi)n \mid n \in \mathcal{J}] = [(xt\varphi)n \mid n \in \mathcal{J}]$, the orbit of x under the discrete flow generated by π^t is not dense in X. The proof is completed.

4.56. REMARK. The only universally valid two-termed implications among the almost periodicity properties are the obvious ones. These implications are summarized in Table 2. (See Part Two.)

4.57. REMARK. Let φ be a function on a topological group T to a uniform space Y. Then the following statements are pairwise equivalent:

(1) φ is {left}{right} uniformly almost periodic.

(2) If α is an index of Y, then there exists a left syndetic subset A of T such that $\tau \in T$ and $t \in A$ implies $\{(\tau\varphi, t\tau\varphi) \in \alpha\}\{(\tau\varphi, \tau t\varphi) \in \alpha\}$.

(3) If α is an index of Y, then there exists a right syndetic subset A of T such that $\tau \in T$ and $t \in A$ implies $\{(\tau\varphi, t\tau\varphi) \in \alpha\}\{(\tau\varphi, \tau t\varphi) \in \alpha\}$.

(4) If α is an index of Y, then there exists a compact subset K of T such that $t \in T$ implies the existence of $k \in K$ such that $\tau \in T$ implies $\{(k\tau\varphi, t\tau\varphi) \in \alpha\}\{(\tau k\varphi, \tau t\varphi) \in \alpha\}$.

TABLE 2

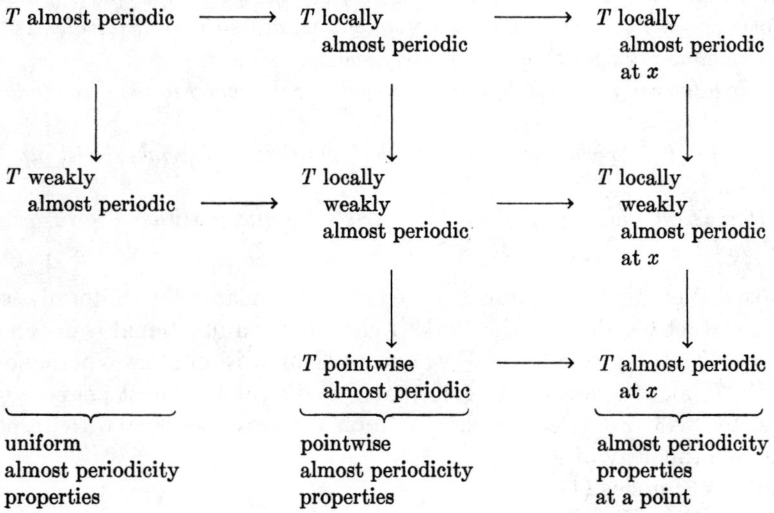
T almost periodic
T locally almost periodic
T locally almost periodic at x
T weakly almost periodic
T locally weakly almost periodic
T locally weakly almost periodic at x
T pointwise almost periodic
T almost periodic at x
uniform almost periodicity properties
pointwise almost periodicity properties
almost periodicity properties at a point

4.58. THEOREM. *Let φ be a continuous $\{$left$\}\{$right$\}$ uniformly almost periodic function on a topological group T to a uniform space Y. Then φ is $\{$left$\}\{$right$\}$ uniformly continuous and bounded.*

PROOF. Suppose φ is continuous and left uniformly almost periodic.

We show φ is left uniformly continuous. Let α be an index of Y. Choose a symmetric index β of Y such that $\beta^3 \subset \alpha$. There exists a compact subset K of T such that $t \in T$ implies the existence of $k \in K$ such that $(k\tau\varphi, t\tau\varphi) \in \beta$ $(\tau \in T)$. Select a neighborhood U of e such that $kU\varphi \subset k\varphi\beta$ $(k \in K)$. It is enough to show that $t \in T$ implies $tU\varphi \subset t\varphi\alpha$. Let $t \in T$. There exists $k \in K$ such that $(k\tau\varphi, t\tau\varphi) \in \beta$ $(\tau \in T)$. Hence $tU\varphi \subset kU\varphi\beta \subset k\varphi\beta^2 \subset t\varphi\beta^3 \subset t\varphi\alpha$ and $tU\varphi \subset t\varphi\alpha$.

We use 4.41 to show that φ is bounded, that is, the range $T\varphi$ of φ is totally bounded. Let α be an index of Y. There exists a compact subset K of T such that $t \in T$ implies the existence of $k \in K$ such that $(k\tau\varphi, t\tau\varphi) \in \alpha$ $(\tau \in T)$. Then $T\varphi \subset K\varphi\alpha$. The proof is completed.

4.59. THEOREM. *Let (Φ, T, ρ) be the $\{$left$\}\{$right$\}$ uniform functional transformation group over a topological group T to a uniform space Y and let $\varphi \in \Phi$. Then:*

(1) *The orbit φT of φ is totally bounded if and only if φ is $\{$left$\}\{$right$\}$ uniformly almost periodic.*

(2) *If Y is complete, then the orbit-closure $\overline{\varphi T}$ of φ is compact if and only if φ is $\{$left$\}\{$right$\}$ uniformly almost periodic.*

PROOF. Use 4.43 and 11.05.

4.60. Remark. Let φ be a complex-valued function on a discrete group. Then φ is {left almost periodic}{right almost periodic}{almost periodic} in the sense of von Neumann if and only if φ is {left}{right}{bilaterally} uniformly almost periodic. (Cf. von Neumann [1].)

4.61. Theorem. *Let φ be a continuous function on a topological group T to a uniform space Y. Then the following statements are pairwise equivalent:*

(1) *φ is bilaterally uniformly almost periodic.*

(2) *φ is bilaterally uniformly almost periodic with respect to the discrete topology of T.*

(3–4) *φ is {left}{right} uniformly almost periodic and {right}{left} uniformly continuous.*

(5) *If α is an index of Y, then there exists a finite partition $\mathcal{E}$ of T such that $E \in \mathcal{E}$ and $t, s \in T$ implies $tEs\varphi \times tEs\varphi \subset \alpha$.*

Proof. We show (1) implies (2). Let φ be bilaterally uniformly almost periodic and let (Φ, T, ρ) be the {left}{right} uniform functional transformation group over T to Y. By 4.58, $\varphi \in \Phi$, and by 3.43, φ is an almost periodic point under (Φ, T, ρ). It follows from 4.43 that T is discretely almost periodic on φT, whence, by 3.43, φ is {left}{right} uniformly almost periodic with respect to the discrete topology of T.

Clearly (2) implies (1).

By 4.58, (1) implies (3) and (4).

We show {(3)}{(4)} implies (5). Assume {(3)}{(4)}. Let (Φ, T, ρ) be the {left} {right} uniform functional transformation group over T to Y. Now $\varphi \in \Phi$; by 3.43, φ is an almost periodic point under (Φ, T, ρ); and by 4.43, φT is totally bounded. Let α be an index of Y and let α_Φ (cf. 11.01) be the corresponding index of Φ. By 4.38, there exists a finite partition $\mathcal{E}$ of T such that $E \in \mathcal{E}$ and $\psi \in \varphi T$ implies $\psi E \times \psi E \subset \alpha_\Phi^*$. Let $E \in \mathcal{E}$, let $t, s \in T$ and let $\psi = \varphi t$. Then $\psi E \times \psi E \subset \alpha_\Phi^*$ implies $tEs\varphi \times tEs\varphi \subset \alpha$. This proves (5).

We show (5) implies (2). Assume (5). Let α be an index of Y and let $\mathcal{E}$ be the corresponding finite partition of T. Let K be a finite subset of T such that $E \in \mathcal{E}$ implies $K \cap E \neq \emptyset$. Let $t \in T$. Then there exists $E \in \mathcal{E}$ and $k \in K$ such that $t, k \in E$. Let $\tau \in T$. It follows that $E\tau\varphi \times E\tau\varphi \subset \alpha$, whence $(k\tau\varphi, t\tau\varphi) \in \alpha$, and $\tau E\varphi \times \tau E\varphi \subset \alpha$, whence $(\tau k\varphi, \tau t\varphi) \in \alpha$. By 4.57, φ is bilaterally uniformly almost periodic with respect to the discrete topology of T.

The proof is completed.

4.62. Remark. Let φ be a continuous function on a topological group T to a uniform space Y and let the left and right uniformities of T coincide. Then φ is left uniformly almost periodic if and only if φ is right uniformly almost periodic. In such an event the words "left" and "right" may be omitted.

4.63. Theorem. *Let T be a topological group, let Y be a uniform space, and let Φ be a finite class of continuous bilaterally uniformly almost periodic functions on T to Y. Then the function class Φ is bilaterally uniformly almost periodic.*

PROOF. Let (Ψ, T, ρ) be the {left} {right} uniform functional transformation group over T to Y. By 4.61, $\Phi \subset \Psi$. Hence, $\Phi T = \bigcup_{\varphi \in \Phi} \varphi T$ is totally bounded. by 4.59. By 4.38, (Ψ, T, ρ) is almost periodic on ΦT and therefore, by 3.45, the function class ΦT, which contains Φ, is {left} {right} uniformly almost periodic. The proof is completed.

4.64. THEOREM. *Let T be a topological group, let Y be a complete uniform space, let n be a positive integer, let O be a continuous n-ary operation in Y, and let $\varphi_1, \cdots, \varphi_n$ be continuous bilaterally uniformly almost periodic functions on T to Y. Then $O(\varphi_1, \cdots, \varphi_n)$ is a continuous bilaterally uniformly almost periodic function on T to Y.*

PROOF. Let α be an index of Y. Now $E = \overline{T\varphi_1} \times \cdots \times \overline{T\varphi_n}$ is compact by 4.59 and therefore O is uniformly continuous on E. Hence, there exists an index β of Y such that $x_i, y_i \in \overline{T\varphi_i}$ $(i = 1, \cdots, n)$ with $(x_i, y_i) \in \beta$ $(i = 1, 2, \cdots, n)$ implies $(O(x_1, \cdots, x_n), O(y_1, \cdots, y_n)) \in \alpha$. By 4.63, there exists a left syndetic subset A of T such that $\tau \in T$ and $t \in A$ implies $\{(\tau\varphi_i, t\tau\varphi_i) \in \beta$ $(i = 1, \cdots, n)\}$ $\{(\tau\varphi_i, \tau t\varphi_i) \in \beta$ $(i = 1, \cdots, n)\}$. Hence, $\tau \in T$ and $t \in A$ implies $\{(O(\tau\varphi_1, \cdots, \tau\varphi_n), O(t\tau\varphi_1, \cdots, t\tau\varphi_n)) \in \alpha\}$ $\{(O(\tau\varphi_1, \cdots, \tau\varphi_n), O(\tau t\varphi_1, \cdots, \tau t\varphi_n)) \in \alpha\}$. The proof is completed.

4.65. THEOREM. *Let (X, T, π) be a transformation group such that X is a compact uniform space and T is almost periodic, let Y be a uniform space, let φ be a continuous function on X to Y, and let $a \in X$. Then the function $\pi_a\varphi : T \to Y$ is bilaterally uniformly almost periodic.*

PROOF. We show $\pi_a\varphi$ is right uniformly almost periodic. Let α be an index of Y. Choose an index β of X so that $(x, y) \in \beta$ implies $(x\varphi, y\varphi) \in \alpha$. There exists a left syndetic subset A of T such that $xA \subset x\beta$ for all $x \in X$. If $\tau \in T$ and $t \in A$, then $(\tau\pi_a, \tau t\pi_a) = (a\tau, a\tau t) \in \beta$ and $(\tau\pi_a\varphi, \tau t\pi_a\varphi) \in \alpha$.

By 4.39, π_a is left uniformly continuous. Hence, $\pi_a\varphi$ is left uniformly continuous. By 4.61, $\pi_a\varphi$ is also left uniformly almost periodic. The proof is completed.

4.66. LEMMA. *Let X be a compact uniform space, let α be an index of X, and let Y be a uniform space which contains an arc. Then there exists an index β of Y and a finite set Φ of continuous functions on X to Y such that $x_1, x_2 \in X$ with $(x_1\varphi, x_2\varphi) \in \beta$ $(\varphi \in \Phi)$ implies $(x_1, x_2) \in \alpha$.*

PROOF. Let E be an arc in Y, let y_0, y_1 be the endpoints of E, and let β be an index of Y such that $(y_0, y_1) \notin \beta$. Choose a symmetric open index γ of X such that $\gamma^3 \subset \alpha$. Select a finite subset A of X such that $X = \bigcup_{a \in A} a\gamma$. Since X is normal, for each $a \in A$ there exists a continuous function φ_a on X to E such that $x\varphi_a = y_0$ $(x \in a\gamma)$ and $x\varphi_a = y_1$ $(x \in X - a\gamma^2)$. Define $\Phi = [\varphi_a \mid a \in A]$. Let $x_1, x_2 \in X$ with $(x_1\varphi, x_2\varphi) \in \beta$ $(\varphi \in \Phi)$. Assume $(x_1, x_2) \notin \alpha$. Now $x_1 \in a\gamma$ for some $a \in A$. It follows that $x_2 \notin a\gamma^2$ since otherwise $x_2 \in a\gamma^2$,

$a \in x_1\gamma$, $x_2 \in x_1\gamma^3$ and $(x_1, x_2) \in \gamma^3 \subset \alpha$. Hence $x_1\varphi_a = y_0$, $x_2\varphi_a = y_1$ and $(y_0, y_1) = (x_1\varphi_a, x_2\varphi_a) \in \beta$. This is a contradiction. The proof is completed.

4.67. THEOREM. *Let (X, T, π) be a transformation group such that X is a compact uniform space which is minimal under T and let Y be a uniform space which contains an arc. Then the following statements are pairwise equivalent:*

(1) *T is almost periodic.*

(2) *If φ is a continuous function on X to Y and if $a \in X$, then the function $\pi_a\varphi : T \to Y$ is bilaterally uniformly almost periodic.*

(3) *If φ is a continuous function on X to Y, then there exists $a \in X$ such that the function $\pi_a\varphi : T \to Y$ is right uniformly almost periodic.*

PROOF. By 4.65, (1) implies (2). Clearly, (2) implies (3). Assume (3). We prove (1).

We first show (I): If φ is a continuous function on X to Y and if $x \in X$, then $\pi_x\varphi$ is a right uniformly almost periodic function.

Let φ be a continuous function on X to Y. By hypothesis, there exists $a \in X$ such that $\pi_a\varphi$ is a right uniformly almost periodic function. Let α be a closed index of Y. There exists a left syndetic subset A of T such that $(a\tau\varphi, a\tau t\varphi) = (\tau\pi_a\varphi, \tau t\pi_a\varphi) \in \alpha$ $(\tau \in T, t \in A)$. By 3.32, $(x\varphi, xt\varphi) \in \alpha$ $(x \in X, t \in A)$. Hence $(\tau\pi_x\varphi, \tau t\pi_x\varphi) = (x\tau\varphi, x\tau t\varphi) \in \alpha$ $(x \in X, \tau \in T, t \in A)$. This proves (I).

Let α be a closed index of X. By 4.66, there exists an index β of Y and a finite set Φ of continuous functions on X to Y such that $x_1, x_2 \in X$ with $(x_1\varphi, x_2\varphi) \in \beta$ $(\varphi \in \Phi)$ implies $(x_1, x_2) \in \alpha$. Choose $a \in X$. By 4.39 and (I), each $\pi_a\varphi$ $(\varphi \in \Phi)$ is a left uniformly continuous right uniformly almost periodic function. By 4.63, the function class $\{\pi_a\varphi \mid \varphi \in \Phi\}$ is right uniformly almost periodic. Hence, there exists a left syndetic subset A of T such that $(a\tau\varphi, a\tau t\varphi) = (\tau\pi_a\varphi, \tau t\pi_a\varphi) \in \beta$ $(\tau \in T, t \in A, \varphi \in \Phi)$. It follows that $(a\tau, a\tau t) \in \alpha$ $(\tau \in T, t \in A)$ and, by 3.32, $(x, xt) \in \alpha$ $(x \in X, t \in A)$.

This proves (1). The proof of the theorem is completed.

4.68. DEFINITION. Let T be a group, let Y be a uniform space, and let $\varphi : T \to Y$. The map φ is said to be *stable* provided that if α is an index of Y, then there exists an index β of Y such that $t, s \in T$ with $(t\varphi, s\varphi) \in \beta$ implies $(t\tau\varphi, s\tau\varphi) \in \alpha$ $(\tau \in T)$ and $(\tau t\varphi, \tau s\varphi) \in \alpha$ $(\tau \in T)$.

4.69. THEOREM. *Let T be a group, let Y be a topological group provided with its bilateral uniformity, and let φ be a homomorphism of T into Y. Then φ is stable.*

PROOF. Let U be a neighborhood of the identity of Y. If $t, s, \tau \in T$ and if $t\varphi \cdot (s\varphi)^{-1} \in U$ and $(t\varphi)^{-1} \cdot s\varphi \in U$, then $t\tau\varphi \cdot (s\tau\varphi)^{-1} \in U$ and $(\tau t\varphi)^{-1} \cdot \tau s\varphi \in U$. The proof is completed.

4.70. REMARK. Let X be a set, let $\circ$ be a group structure of X, let $\mathfrak{U}$ be a uniformity of X, let $\mathfrak{I}$ be the topology of X induced by $\mathfrak{U}$, and for each $\alpha \in \mathfrak{U}$ suppose there exists $\beta \in \mathfrak{U}$ such that $(x, y) \in \beta$ implies $(z \circ x, z \circ y) \in \alpha$ $(z \in X)$

and $(x \circ z, y \circ z) \in \alpha$ $(z \in X)$. Then $(X, \circ, \ )$ is a topological group such that the bilateral uniformity of $(X, \circ, \mathfrak{J})$ coincides with $\mathfrak{U}$.

4.71. THEOREM. *Let T be a group, let Y be a separated uniform space, and let φ be a map of T onto Y. Then φ is stable if and only if there exists a (necessarily unique) group structure $\circ$ of Y such that $(Y, \circ)$ is a topological group, $\varphi : T \to (Y, \circ)$ is a homomorphism, and the bilateral uniformity of $(Y, \circ)$ coincides with the given uniformity of Y.*

PROOF. The necessity follows from 4.69. We prove the sufficiency. Suppose φ is stable. Define $G = e\varphi\varphi^{-1}$.

We observe that if $t, s \in T$, then the following statements are pairwise equivalent: $t\varphi = s\varphi$; $t\tau\varphi = s\tau\varphi$ $(\tau \in T)$; $\tau t\varphi = \tau s\varphi$ $(\tau \in T)$; $ts^{-1} \in G$.

Now G is a group since $e \in G$ and from $t, s \in G$ it follows that $t\varphi = e\varphi = s\varphi$ and $ts^{-1} \in G$. Also G is invariant in T since from $t \in G$ and $\tau \in T$ it follows that $t\varphi = e\varphi$, $\tau t\varphi = \tau e\varphi = \tau\varphi$, $\tau t\tau^{-1} = \tau\tau^{-1}\varphi = e\varphi$, and $\tau t\tau^{-1} \in G$. Since φ is constant on each translate of G, there exists a map $\hat{\varphi} : T/G \to Y$ such that $\pi\hat{\varphi} = \varphi$ where π is the projection of T onto T/G. Now $\hat{\varphi}$ is a one-to-one map of T/G onto Y. Let $\circ$ be the unique group structure of Y such that $\hat{\varphi}$ is an isomorphism of T/G onto $(Y, \circ)$. Clearly, $\varphi = \pi\hat{\varphi}$ is a homomorphism of T onto $(Y, \circ)$.

To finish the proof we use 4.70. Let α be an index of Y. There exists an index β of Y such that $t, s \in T$ with $(t\varphi, s\varphi) \in \beta$ implies $(t\varphi \circ \tau\varphi, s\varphi \circ \tau\varphi) = (t\tau\varphi, s\tau\varphi) \in \alpha$ $(\tau \in T)$ and $(\tau\varphi \circ t\varphi, \tau\varphi \circ s\varphi) = (\tau t\varphi, \tau s\varphi) \in \alpha$ $(\tau \in T)$. The proof is completed.

4.72. THEOREM. *Let T be a discrete group, let Y be a topological group provided with its bilateral uniformity, and let φ be a homomorphism of T onto Y. Then φ is a uniformly almost periodic function if and only if Y is totally bounded.*

PROOF. Use 4.57 and 4.69.

4.73. THEOREM. *Let T be a discrete group, let Y be a totally bounded separated uniform space, and let φ be a stable map of T into Y. Then φ is a uniformly almost periodic function.*

PROOF. Use 4.71 and 4.72.

4.74. STANDING NOTATION. For the remainder of this section we adopt the following notation.

Let N be the set of all positive integers. Let X be a set. If $n \in N$, then X^n or $X^{(n)}$ denotes the nth cartesian power of X. The *total power of X*, denoted X^*, is $\bigcup_{n \in N} X^n$. The phrase "finite family" shall mean "nonvacuous finite family". We consider any finite ordered family in X to be an element of X^*. We define the binary operation *composition* in X^* : if $a = (x_1, \cdots, x_n) \in X^*$ and if $b = (y_1, \cdots, y_m) \in X^*$, then $ab = (x_1, \cdots, x_n, y_1, \cdots, y_m)$. Composition is associative but not commutative. If $n \in N$ and if $x \in X$, then x^n or $x^{(n)}$ denotes the n-tuple $(x, \cdots, x)$. If $(a_\iota \mid \iota \in I)$ is a finite ordered family in X^*, then

$\prod_{\iota \in I} a_\iota$ denotes the continued composition $a_{\iota_1} \cdots a_{\iota_n}$ where $I = [\iota_1, \cdots, \iota_n]$ and $\iota_1 < \cdots < \iota_n$. If $\alpha \subset X \times X$, then α^* denotes the set of all couples (a, b) such that for some $n \in N$ we have $a = (x_1, \cdots, x_n) \in X^n$, $b = (y_1, \cdots, y_n) \in X^n$ and $(x_\iota, y_\iota) \in \alpha$ $(\iota = 1, \cdots, n)$.

4.75. Definition. Let X be a uniform space. An *averaging process in X* is defined to be a function $\mu : X^* \to X$ such that:

(1) If $x \in X$ and if $n \in N$, then $x^n\mu = x$.

(2) If $a \in X^*$ and if b is a permutation of a, then $a\mu = b\mu$.

(3) If $n \in N$ and if $(a_\iota \mid \iota \in I)$ is a finite ordered family in X^n, then $(a_\iota\mu \mid \iota \in I)\mu = (\prod_{\iota \in I} a_\iota)\mu$.

(4) If α is an index of X, then there exists an index β of X such that $(a, b) \in \beta^*$ implies $(a\mu, b\mu) \in \alpha$.

4.76. Standing notation. Let X be a uniform space, let μ be an averaging process in X, let T be a discrete group and let $\varphi : T \to X$.

4.77. **Remark.** **By 4.75(2), if $(x_\iota \mid \iota \in I)$ is a finite family in X, then $(x_\iota \mid \iota \in I)\mu$ is uniquely defined. Similarly, if E is a finite subset of X, then $E\mu$ is uniquely defined.**

Let α and β be indices of X such that $(a, b) \in \beta^*$ implies $(a\mu, b\mu) \in \alpha$. It follows from 4.75(1) that if $x \in X$ and if $(x_\iota \mid \iota \in I)$ is a finite family in $x\beta$, then $(x_\iota \mid \iota \in I)\mu \in x\alpha$.

4.78. Definition. Let α be an index of X. An *α-mean of φ* is a point x of X such that $tas\varphi\mu \in x\alpha$ $(t, s \in T)$ for some $a \in T^*$. The set of all α-means of φ is denoted by $\alpha\varphi$.

4.79. Remark. If α and β are indices of X such that $\beta \subset \alpha$, then $\beta\varphi \subset \alpha\varphi$. If α and β are indices of X, then $(\alpha \cap \beta)\varphi \subset \alpha\varphi \cap \beta\varphi$.

4.80. Lemma. *Let α be an index of X. Then:*

(1) *There exists an index β of X such that $\overline{\beta\varphi} \subset \alpha\varphi$.*

(2) *There exists an index β of X such that $\beta\varphi \times \beta\varphi \subset \alpha$.*

Proof. (1) Let β be an index of X such that $\beta^2 \subset \alpha$. Let $x \in \overline{\beta\varphi}$. Choose $y \in \beta\varphi \cap x\beta$. There exists $a \in T^*$ such that $tas\varphi\mu \in y\beta$ $(t, s \in T)$. Since $y\beta \subset x\beta^2 \subset x\alpha$, we have $tas\varphi\mu \in x\alpha$ $(t, s \in T)$ and $x \in \alpha\varphi$. The proof is completed.

(2) Let γ be a symmetric index of X such that $\gamma^2 \subset \alpha$. Let β be an index of X such that if $x \in X$ and if $(x_\lambda \mid \lambda \in \Lambda)$ is a finite family in $x\beta$, then $(x_\lambda \mid \lambda \in \Lambda)\mu \in x\gamma$. Let $x, y \in \beta\varphi$. There exists $a, b \in T^*$ such that $tas\varphi\mu \in x\beta$ $(t, s \in T)$ and $tbs\varphi\mu \in y\beta$ $(t, s \in T)$. Let $a = (\tau_\iota \mid \iota \in I)$ and let $b = (\sigma_\kappa \mid \kappa \in K)$. Now $a\sigma_\kappa\varphi\mu \in x\beta$ $(\kappa \in K)$ and $\tau_\iota b\varphi\mu \in y\beta$ $(\iota \in I)$. It follows that $(\prod_{\kappa \in K} a\sigma_\kappa)\varphi\mu = (\prod_{\kappa \in K} a\sigma_\kappa\varphi)\mu = (a\sigma_\kappa\varphi\mu \mid \kappa \in K)\mu \in x\gamma$ and $(\prod_{\iota \in I} \tau_\iota b)\varphi\mu = (\prod_{\iota \in I} \tau_\iota b\varphi)\mu = (\tau_\iota b\varphi\mu \mid \iota \in I)\mu \in y\gamma$. Since $\prod_{\kappa \in K} a\sigma_\kappa$ and $\prod_{\iota \in I} \tau_\iota b$ are permutations of each other, $(\prod_{\kappa \in K} a\sigma_\kappa)\varphi\mu = (\prod_{\iota \in I} \tau_\iota b)\varphi\mu$ and we conclude that $x\gamma \cap y\gamma \neq \emptyset$ whence $(x, y) \in \gamma^2 \subset \alpha$. The proof is completed.

4.81. Definition. Let $x \in X$. The point x is said to be a *mean of* φ provided that if α is an index of X, then there exists $a \in T^*$ such that $t, s \in T$ implies $tas\varphi\mu \in x\alpha$.

4.82. Remark. Let $x \in X$. Then x is a mean of φ if and only if x is an α-mean of φ for each index α of X.

4.83. Lemma. *Let X be complete. Then there exists a mean of φ if and only if there exists an α-mean of φ for each index α of X.*

Proof. Let $\mathfrak{U}$ be the uniformity of X. Assume $\alpha\varphi \neq \emptyset$ $(\alpha \in \mathfrak{U})$. By 4.79 and 4.80(2), $[\alpha\varphi \mid \alpha \in \mathfrak{U}]$ is a cauchy filter base on X. Since $\bigcap_{\alpha \in \mathfrak{U}} \alpha\varphi = \bigcap_{\alpha \in \mathfrak{U}} \alpha\varphi$ by 4.80(1), it follows that $\bigcap_{\alpha \in \mathfrak{U}} \alpha\varphi \neq \emptyset$. The converse is obvious. The proof is completed.

4.84. Lemma. *Let $(X_\iota \mid \iota \in I)$ be a finite family of sets. Then there exists a one-to-one choice function of $(X_\iota \mid \iota \in I)$ if and only if $K \subset I$ implies* crd $K \leqq$ crd $\bigcup_{\kappa \in K} X_\kappa$.

Proof. The necessity is obvious. We prove the sufficiency. The theorem is true when crd $I = 0$ or 1. Let n be an integer such that $n > 1$. Assume the theorem is true when crd $I < n$. We show the theorem is true when crd $I = n$. Suppose crd $I = n$. If $K \subset I$ with $0 <$ crd $K < n$ implies crd $K <$ crd $\bigcup_{\kappa \in K} X_\kappa$, then choosing $\mu \in I$ and $x \in X_\mu$ it follows that $\Lambda \subset I - \mu$ implies crd $\Lambda \leqq \bigcup_{\lambda \in \Lambda} (X_\lambda - x)$, whence $(X_\iota - x \mid \iota \in I - \mu)$ has a one-to-one choice function by the induction assumption and the proof is completed. Suppose, on the contrary, there exists $K \subset I$ such that $0 <$ crd $K < n$ and crd $K =$ crd X where $X = \bigcup_{\kappa \in K} X_\kappa$. By the induction assumption $(X_\kappa \mid \kappa \in K)$ has a one-to-one choice function. It is therefore enough to show $(X_\iota - X \mid \iota \in I - K)$ has a one-to-one choice function. To do this it suffices by the induction assumption to show that $\Lambda \subset I - K$ implies crd $\Lambda \leqq$ crd $\bigcup_{\lambda \in \Lambda} (X_\lambda - X)$. However, if $\Lambda \subset I - K$ with crd $\Lambda >$ crd $\bigcup_{\lambda \in \Lambda} (X_\lambda - X)$, then crd $(K \cup \Lambda) =$ crd $K +$ crd $\Lambda >$ crd $X +$ crd $\bigcup_{\lambda \in \Lambda} (X_\lambda - X) \geqq$ crd $\bigcup_{\mu \in K \cup \Lambda} X_\mu$ which contradicts the hypothesis. The proof is completed.

4.85. Lemma. *Let X be a set and let $\mathcal{E}$, $\mathfrak{F}$ be partitions of X with the same finite cardinal. Then the following statements are pairwise equivalent:*

(1) $\mathfrak{A} \subset \mathcal{E}$, $\mathfrak{B} \subset \mathfrak{F}$ *and* $\bigcup \mathfrak{A} \subset \bigcup \mathfrak{B}$ *implies* crd $\mathfrak{A} \leqq$ crd $\mathfrak{B}$.

(2) $\mathfrak{A} \subset \mathcal{E}$, $\mathfrak{B} \subset \mathfrak{F}$ *and* $\bigcup \mathfrak{B} \subset \bigcup \mathfrak{A}$ *implies* crd $\mathfrak{B} \leqq$ crd $\mathfrak{A}$.

(3) *There exists a common choice set of* $\mathcal{E}$ *and* $\mathfrak{F}$.

Proof. Use 4.84.

4.86. Theorem. *Let X be complete and let φ be uniformly almost periodic. Then there exists a mean of φ.*

Proof. Let α be an index of X. It is enough by 4.83 to show that $\alpha\varphi \neq \emptyset$. Let β be an index of X such that $\beta^2 \subset \alpha$. Choose an index γ of X such that

$(c, d) \in \gamma^*$ implies $(c\mu, d\mu) \in \beta$. By 4.61(5) there exists a finite partition $\mathcal{E}$ of T with least cardinal such that $tEs\varphi \times tEs\varphi \subset \gamma$ $(t, s \in T; E \in \mathcal{E})$. We observe that if $p = (p_\iota \mid \iota \in I)$ and $q = (q_\iota \mid \iota \in I)$ are choice functions of $\mathcal{E}$, then $(tps\varphi\mu, tqs\varphi\mu) \in \beta$ $(t, s \in T)$; for we may suppose that for each $\iota \in I$ the points p_ι and q_ι belong to a common member of $\mathcal{E}$, whence $t, s \in T$ implies $(tp_\iota s\varphi, tq_\iota s\varphi) \in \gamma$ $(\iota \in I)$, $(tps\varphi, tqs\varphi) \in \gamma^*$ and $(tps\varphi\mu, tqs\varphi\mu) \in \beta$. Let $a = (\tau_\iota \mid \iota \in I)$ be a choice function of $\mathcal{E}$ and define $x = a\varphi\mu$. We show $x \in \alpha\varphi$. Let $t, s \in T$. By 4.85 there exists a common choice function $b = (\sigma_\iota \mid \iota \in I)$ of $\mathcal{E}$ and $t^{-1}\mathcal{E}s^{-1}$. Since a and tbs are choice functions of $\mathcal{E}$, we have $(a\varphi\mu, tbs\varphi\mu) \in \beta$. Since b and a are choice functions of $\mathcal{E}$, we have $(tbs\varphi\mu, tas\varphi\mu) \in \beta$. Thus $(a\varphi\mu, tas\varphi\mu) \in \beta^2 \subset \alpha$ and $tas\varphi\mu \in x\alpha$. The proof is completed.

4.87. Notes and references.

A number of the results of the early part of this section can be found in Gottschalk [3, 6, 7].

(4.02) It was observed later that if x is almost periodic then x is discretely almost periodic.

(4.48) For related results, see Stepanoff and Tychonoff [1].

(4.52 and 4.53) Cf. Halmos and Samelson [1].

(4.55) This theorem is due to E. E. Floyd (Personal communication).

(4.59) The connection between compactness and almost periodicity of functions was first observed by Bochner (cf. Bochner [1]).

(4.61(5)) This characterization of an almost periodic function is due to W. Maak (cf. Maak [1, 2])

(4.65 and 4.67) Forms of these theorems were originally proved by J. D. Baum (cf. Baum [1]).

(4.73) The connections between stability and almost periodicity of functions have been observed and studied many times (cf. Franklin [1], Markoff [2], Bohr [1, C32], Hartman and Wintner [1]).

(4.85) Cf. Halmos and Vaughan [3].

5. REGULAR ALMOST PERIODICITY

5.01. STANDING NOTATION. Throughout this section (X, T, π) denotes a transformation group.

5.02. REMARK. Let $x \in X$. Then:

(1) If T is regularly almost periodic at x, then T is isochronous at x.

(2) If T is isochronous at x, then T is almost periodic at x.

5.03. THEOREM. *Let $x \in X$ and let T be {regularly almost periodic}{isochronous} at x. Then T is {regularly almost periodic}{isochronous} at every point of xT.*

PROOF. Use 3.21.

5.04. LEMMA. *Let T be a discrete group. Then:*

(1) *If G is a syndetic subgroup of T, then there exists a syndetic invariant subgroup H of T such that $H \subset G$.*

(2) *If G and H are syndetic subgroups of T, then $G \cap H$ is a syndetic subgroup of T, of G, of H.*

PROOF. (1) For $t \in T$ let $\varphi_t : T/G \to T/G$ be the permutation of T/G defined by $E\varphi_t = t^{-1}E$ $(E \in T/G)$. Let P be the permutation group of T/G. Define the homomorphism $\varphi : T \to P$ by $t\varphi = \varphi_t$ $(t \in T)$. Define H to be the kernel of φ. The conclusion follows.

(2) If G and H are subgroups of T, then $t(G \cap H) = tG \cap tH$ $(t \in T)$ whence $T/G \cap H \subset T/G \mathbin{\dot\cap} T/H$. The conclusion follows.

5.05. REMARK. Let T be discrete and let $x \in X$. Then the following statements are equivalent:

(1) T is regularly almost periodic at x.

(2) If U is a neighborhood of x, then there exists a syndetic subgroup A of T such that $xA \subset U$.

5.06. REMARK. Let T be discrete and let $x \in X$. Then the following statements are equivalent:

(1) T is isochronous at x.

(2–3) If U is a neighborhood of x, then there exist a syndetic subgroup A of T and $t \in T$ such that $\{xtA \subset U\}\{xAt \subset U\}$.

5.07. INHERITANCE THEOREM. *Let T be discrete and let S be a syndetic subgroup of T. Then:*

(1) *If $x \in X$, then S is {regularly almost periodic}{isochronous} at x if and only if T is {regularly almost periodic}{isochronous} at x.*

(2) *S is pointwise {regularly almost periodic} {isochronous} if and only if T is pointwise {regularly almost periodic} {isochronous}.*

PROOF. From 5.04 the first reading is obvious.

Let $x \in X$. To prove the second reading it is enough to show that if T is isochronous at x, then S is isochronous at x. By 5.04(1) we assume without loss that S is also invariant in T.

We first show that if T is isochronous at x, then S_x is isochronous at x. Suppose T is isochronous at x. Let U be a neighborhood of x. By 2.10(2), $x \notin \overline{x(T - S_x)}$. Hence we may assume that $U \cap x(T - S_x) = \emptyset$. There exist $t \in T$ and a syndetic subgroup A of T such that $xtA \subset U$. Therefore $xtA \cap x(T - S_x) = \emptyset$, $tA \subset S_x$, $t \in S_x$ and $A \subset t^{-1}S_x = S_x$ by 2.10(1). Thus $t \in S_x$ and A is a syndetic subgroup of S_x such that $xtA \subset U$. This shows that S_x is isochronous at x.

We next show that if S_x is isochronous at x, then S is isochronous at x. Suppose S_x is isochronous at x. Let U be an open neighborhood of x. By 2.10(3) there exists a finite subset M of T such that $xM \subset U$ and $S_x \subset SM^{-1}$. Let V be a neighborhood of x for which $VM \subset U$. There exist $t \in S_x$ and a syndetic subgroup A of S_x such that A is invariant in T and $xtA \subset V$. Choose $s \in S$ and $m \in M$ such that $t = sm^{-1}$. Define $B = S \cap A$. Then $xsB = xsm^{-1}Bm \subset xtAM \subset VM \subset U$. Thus $s \in S$ and B is a syndetic subgroup of S such that $xsB \subset U$. This shows that S is isochronous at x. The proof is completed.

5.08. THEOREM. *Let X be regular, let T be discrete and let T be pointwise regularly almost periodic. Then every orbit-closure under T is zero-dimensional.*

PROOF. Let $x \in X$. We assume without loss that $X = \overline{xT}$. Let U be a neighborhood of x. There exists a subgroup A of T and a finite subset E of T such that $T = AE$ and $\overline{xA} \subset U$. By 5.07(2), A is pointwise regularly almost periodic. By 4.08, the class $\mathcal{C}$ of all orbit-closures under A is a partition of X. Since $X = \overline{xT} = \overline{xAE}$, $\mathcal{C}$ is finite. Hence $\overline{xA}$ is an open-closed neighborhood of x. The proof is completed.

5.09. THEOREM. *Let X be a metrizable minimal orbit-closure under T, let T be discrete, let R be the set of all points of X at which T is regularly almost periodic and let $R \neq \emptyset$. Then R is a T-invariant residual G_δ subset of X.*

PROOF. Clearly, R is T-invariant. Let $\mathcal{V}$ be a countable base of the uniformity of X such that every element of $\mathcal{V}$ is a closed index and let $\mathcal{B}$ be the class of all syndetic invariant subgroups of T. We use the notation of 3.30. If $B \in \mathcal{B}$, then by 2.25 the class of all orbit-closures under B is a finite partition of X. It follows that if $\beta \in \mathcal{V}$ and if $B \in \mathcal{B}$, then $E(B, \beta)$ is a union of orbit-closures under B and hence $E(B, \beta)$ is open in X. Since $R = \bigcap_{\beta \in \mathcal{V}} \bigcup_{B \in \mathcal{B}} E(B, \beta)$ and $\mathcal{V}$ is countable, we conclude that R is a residual G_δ subset of X inasmuch as R is represented as a countable intersection of everywhere dense open sets.

5.10. STANDING NOTATION. For the remainder of this section X denotes a uniform space.

5.11. REMARK. The following statements are valid:
(1) T is regularly almost periodic if and only if T is isochronous.
(2) If T is isochronous, then T is almost periodic.

5.12. THEOREM. *Let Y be a T-invariant subset of X and let T be regularly almost periodic on Y. Then T is regularly almost periodic on $\overline{Y}$.*

PROOF. Use 3.33.

5.13. REMARK. Let T be discrete. Then the following statements are equivalent:
(1) T is regularly almost periodic.
(2) If α is an index of X, then there exists a syndetic subgroup A of T such that $x \in X$ implies $xA \subset x\alpha$.

5.14. INHERITANCE THEOREM. *Let T be discrete and let S be a syndetic subgroup of T. Then S is regularly almost periodic if and only if T is regularly almost periodic.*

PROOF. Use 5.04.

5.15. THEOREM. *Let X be locally compact, let $x \in X$, let T be isochronous at x and let T be equicontinuous. Then T is regularly almost periodic on $\overline{xT}$.*

PROOF. By 4.09 we may suppose that X is compact. By 5.12 it is enough to show that T is regularly almost periodic on xT. Let α be an index of X. Choose a symmetric index β of X such that $\beta^2 \subset \alpha$. There exists an index γ of X such that $y \in X$ and $t \in T$ implies $y\gamma t \subset yt\beta$. There exist a syndetic invariant subgroup A of T and $s \in T$ such that $xsA \subset x\gamma$. If $t \in T$, then $xtA = xsAs^{-1}t \subset x\gamma s^{-1}t \subset xs^{-1}t\beta$, $xtA \times xtA \subset \beta^2$ and $xtA \subset xt\beta^2 \subset xt\alpha$. Thus $t \in T$ implies $xtA \subset xt\alpha$. The proof is completed.

5.16. LEMMA. *Let X be compact and suppose that if $x, y \in X$ with $x \neq y$, then there exist a neighborhood U of x, a neighborhood V of y and an index α of X such that $t \in T$ implies $(Ut \times Vt) \cap \alpha = \emptyset$. Then T is uniformly equicontinuous.*

PROOF. For $A \subset X \times X$ and $B \subset T$, define $AB = [(xt, yt) \mid (x, y) \in A, t \in B]$.

Let β be an open index of X. For each $z \in X \times X - \beta$ there exists a neighborhood W_z of z and an index α_z of X such that $W_zT \cap \alpha_z = \emptyset$, whence $\alpha_zT \subset X \times X - W_z$. Choose a finite subset E of $X \times X$ such that $X \times X - \beta \subset \bigcup_{z\in E} W_z$, whence $\bigcap_{z\in E} (X \times X - W_z) \subset \beta$. Define $\alpha = \bigcap_{z\in E} \alpha_z$. Now α is an index of X. Since $\alpha T \subset \bigcap_{z\in E} \alpha_zT \subset \bigcap_{z\in E} (X \times X - W_z) \subset \beta$, it follows that $\alpha T \subset \beta$. The proof is completed.

5.17. THEOREM. *Let X be compact. Consider the following statements:*
(I) T is pointwise isochronous and equicontinuous.

(II) *T is pointwise regularly almost periodic and S is weakly almost periodic for every syndetic invariant subgroup S of T.*

(III) *T is regularly almost periodic.*

Then:

(1) I *is equivalent to* II; III *implies* I *and* II.

(2) *If T is discrete, then* I, II, III *are pairwise equivalent.*

PROOF. By 5.15 and 4.38, I implies II.

Assume II. We prove I. Clearly, T is pointwise isochronous. It remains to prove that T is equicontinuous. Let $x, y \in X$ with $x \neq y$. By 5.16, it is enough to show that there exist a neighborhood U of x, a neighborhood V of y and an index α of X such that $(U \times V)T \cap \alpha = \emptyset$. Since T is regularly almost periodic at x, there exists a syndetic invariant subgroup S of T such that $y \notin \overline{xS}$. By 4.24 the class $\mathcal{A}$ of all orbit-closures under S is a star-closed decomposition of X. By 2.30 and 2.37, $\mathcal{A}$ is star-indexed. Since $y \notin \overline{xS}$, we have $\overline{xS} \cap \overline{yS} = \emptyset$, that is, $x\mathcal{A} \cap y\mathcal{A} = \emptyset$. Choose an index β of X such that $x\mathcal{A}\beta \cap y\mathcal{A}\beta = \emptyset$. Let γ be an index of X such that $\gamma^2 \subset \beta$. Since $\mathcal{A}$ is star-indexed, there exists an index δ of X such that $x\delta\mathcal{A} \subset x\mathcal{A}\gamma$ and $y\delta\mathcal{A} \subset y\mathcal{A}\gamma$. Provide $\mathcal{A}$ with its partition uniformity, which induces its partition topology by 2.36. Clearly, $\mathcal{A}$ is compact. Let $(\mathcal{A}, T, \rho)$ be the partition transformation group of $\mathcal{A}$ induced by (X, T, π). Since $AS = A$ $(A \in \mathcal{A})$, it follows that $(\mathcal{A}, T, \rho)$ is periodic and thus almost periodic. By 4.35, $(\mathcal{A}, T, \rho)$ is discretely almost periodic and hence, by 4.38, $(\mathcal{A}, T, \rho)$ is uniformly equicontinuous. If μ is an index of X, then $\mu^* = [(A, B) \mid A, B \in \mathcal{A}, A \subset B\mu, B \subset A\mu]$ is an index of $\mathcal{A}$. Since $(\mathcal{A}, T, \rho)$ is uniformly equicontinuous, there exists an index θ of $\mathcal{A}$ such that $(A, B) \in \theta$ and $t \in T$ implies $(At, Bt) \in \gamma^*$. Since the projection of X onto $\mathcal{A}$ is uniformly continuous, there exists an index α of X such that $(p, q) \in \alpha$ implies $(p\mathcal{A}, q\mathcal{A}) \in \theta$. Define $U = x\delta$ and $V = y\delta$. Let $x_1 \in U$, $y_1 \in V$ and $t \in T$. We must show $(x_1t, y_1t) \notin \alpha$. Assume $(x_1t, y_1t) \in \alpha$. Then $(x_1\mathcal{A}t, y_1\mathcal{A}t) = (x_1t\mathcal{A}, y_1t\mathcal{A}) \in \theta$, whence $(x_1\mathcal{A}, y_1\mathcal{A}) \in \gamma^*$, $x_1\mathcal{A} \subset y_1\mathcal{A}\gamma$ and $x_1\mathcal{A}\gamma \cap y_1\mathcal{A}\gamma \neq \emptyset$. Now $x_1\mathcal{A}\gamma \subset x\delta\mathcal{A}\gamma \subset x\mathcal{A}\gamma^2 \subset x\mathcal{A}\beta$ and $y_1\mathcal{A}\gamma \subset y\delta\mathcal{A}\gamma \subset y\mathcal{A}\gamma^2 \subset y\mathcal{A}\beta$. Since $x\mathcal{A}\beta \cap y\mathcal{A}\beta = \emptyset$, it follows that $x_1\mathcal{A}\gamma \cap y_1\mathcal{A}\gamma = \emptyset$. This is a contradiction. This proves I.

By 4.37, III implies I.

Now assume T discrete. Assume II. We prove III. Let α be an index of X. Choose a symmetric index β of X such that $\beta^4 \subset \alpha$. For each $x \in X$ there exists a syndetic invariant subgroup A_x of T such that $xA_x \subset x\beta$. By 4.25, for each $x \in X$ there exists a neighborhood U_x of x such that $U_xA_x \subset xA_x\beta$. Select a finite subset E of X for which $X = \bigcup_{x \in E} U_x$. Define $A = \bigcap_{x \in E} A_x$. Then A is an invariant subgroup of T and by 5.04, A is syndetic in T. We show that $x_0 \in X$ implies $x_0A \subset x_0\alpha$. Let $x_0 \in X$. Choose $x \in E$ for which $x_0 \in U_x$. Since $x_0 \in x_0A_x \subset xA_x\beta \subset x\beta^2$ it follows that $x \in x_0\beta^2$ and $x_0A \subset x_0A_x \subset x\beta^2 \subset x_0\beta^4 \subset x_0\alpha$. Hence $x_0A \subset x_0\alpha$. The proof is completed.

5.18. THEOREM. *Let X be a compact minimal orbit-closure under T. Then T is regularly almost periodic if and only if T is pointwise regularly almost periodic.*

PROOF. Use 4.29, 5.15 and 5.17

5.19. THEOREM. *Let X be locally compact separated and let T be pointwise regularly almost periodic. Then T is regularly almost periodic on each orbit-closure under T.*

PROOF. Use 5.18 and 4.09

5.20. REMARK. The following statements are equivalent:

(1) T is weakly isochronous.

(2) If α is an index of X, then there exist a syndetic invariant subgroup A of T and a finite subset K of T such that $x \in X$ implies the existence of $k \in K$ such that $xkA \subset x\alpha$.

5.21. REMARK. Let T be discrete. Then the following statements are equivalent:

(1) T is weakly isochronous.

(2–3) If α is an index of X, then there exist a syndetic subgroup A of T and a finite subset K of T such that $x \in X$ implies the existence of $k \in K$ such that $\{xkA \subset x\alpha\}\{xAk \subset x\alpha\}$.

5.22. THEOREM. *Let Y be a T-invariant subset of X and let T be weakly isochronous on Y. Then T is weakly isochronous on $\overline{Y}$.*

PROOF. Use 3.33.

5.23. THEOREM. *Let X be a compact minimal orbit-closure under T. Consider the following statements:*

(I) *T is regularly almost periodic at some point of X.*

(II) *T is isochronous at some point of X.*

(III) *If α is an index of X, then there exist $x \in X$ and a syndetic invariant subgroup A of T such that $xA \subset x\alpha$.*

(IV) *T is weakly isochronous.*

Then:

(1) I *implies* II; II, III, IV *are pairwise equivalent.*

(2) *If X is metrizable, then* I, II, III, IV *are pairwise equivalent.*

PROOF. (1) Clearly, I implies II; II implies III; IV implies II. Assume III. We prove IV. Let α be an index of X. Choose a symmetric index β of X for which $\beta^3 \subset \alpha$. By 4.21 and 4.24 there exists a finite subset F of T such that $x \in X$ implies $xT \subset x\beta F^{-1}$. Select an index γ of X such that $x \in X$ and $s \in F$ implies $x\gamma s \subset xs\beta$. There exists an index δ of X such that $\delta^2 \subset \gamma$. Choose $x_0 \in X$ and a syndetic invariant subgroup A of T such that $x_0A \subset x_0\delta$. By 4.25 and 4.29 there exists a neighborhood U of x_0 such that $UA \subset x_0A\delta$. Select a finite subset E of T for which $X = UE^{-1}$. Define $K = EF$. We show $x \in X$ implies the existence of $k \in K$ such that $xkA \subset x\alpha$. Let $x \in X$. Choose $t \in E$ such that $xt \in U$ and then choose $s \in F$ such that $xts \in x\beta$. Since $xtAs \subset UAs \subset x_0A\delta s \subset$

$x_0\delta^2 s \subset x_0\gamma s \subset x_0 s\beta$ and $xtAs \cap x\beta \neq 0$, it follows that $xtAs \subset x\beta^3 \subset x\alpha$. Define $k = ts$. Then $k \in K$ and $xkA \subset x\alpha$. The proof of (1) is completed.

(2) Assume III. We prove I.

We first show: (L) If U is a nonvacuous open subset of X, then there exist $x \in U$ and a syndetic invariant subgroup A of T such that $\overline{xA} \subset U$.

Let U be a nonvacuous open subset of X. Choose $x_0 \in U$ and an index α of X such that $\overline{x_0\alpha^2} \subset U$. There exists a finite subset E of T such that $X = x_0\alpha E^{-1}$. Select an index β of X such that $(x_1, x_2) \in \beta$ and $t \in E$ implies $(x_1 t, x_2 t) \in \alpha$. There exists $y \in X$ and a syndetic invariant subgroup A of T such that $\overline{yA \times yA} \subset \beta$. Choose $t \in E$ for which $yt \in x_0\alpha$. Define $x = yt$. Since $\overline{xA \times xA} = \overline{yAt \times yAt} \subset \alpha$ and $xA \cap x_0\alpha \neq \emptyset$, it follows that $\overline{xA} \subset x_0\alpha^2$ and $\overline{xA} \subset \overline{x_0\alpha^2} \subset U$. This proves (L).

Let $[\alpha_n \mid n = 1, 2, \cdots]$ be a countable base of the uniformity of X. Define $V_1 = X$. For $n = 1, 2, \cdots$, we proceed inductively as follows:

There exists a nonvacuous open subset U_n of V_n such that $U_n \times U_n \subset \alpha_n$. By (L) there exist $x_n \in U_n$ and a syndetic invariant subgroup A_n of T such that $\overline{x_n A_n} \subset U_n$. By 2.32 there exists an open neighborhood V_{n+1} of x_n such that $\overline{V_{n+1}} \subset U_n$ and $\overline{V_{n+1} A_n} \subset U_n$.

It is clear that T is regularly almost periodic at every point of $\bigcap_{n=1}^{+\infty} U_n$. Since $\bigcap_{n=1}^{+\infty} U_n \neq \emptyset$, the proof is completed.

5.24. THEOREM. *Let X be a compact metrizable minimal orbit-closure under T and let T be isochronous at some point of X. Then T is locally almost periodic.*

PROOF. Use 4.30 and 5.23.

5.25. THEOREM. *Let X be separated, let $x \in X$, let there exist a compact neighborhood of x and let T be isochronous at x. Then T is weakly isochronous on $\overline{xT}$.*

PROOF. Use 5.02, 4.09, 4.07 and 5.23.

5.26. LEMMA. *Let G be a zero-dimensional locally compact topological group such that the left and right uniformities of G coincide. Then every neighborhood of the identity of G contains an open-closed invariant subgroup of G.*

PROOF. Let U be an open-compact symmetric neighborhood of e. Choose an open-closed symmetric neighborhood V of e such that $UV \subset U$. Let H be the subgroup of G generated by V. Define $K = \bigcap_{x \in G} xHx^{-1}$. Then K is an open-closed invariant subgroup of G such that $K \subset U$.

5.27. LEMMA. *Let X be a compact uniform space, and let Φ be a zero-dimensional compact topological homeomorphism group of X. Then Φ is regularly almost periodic.*

PROOF. Use 5.26.

5.28. LEMMA. *Let G be a discrete group, let H be a topological group with identity e and let φ be a homomorphism of G into H such that $\overline{G\varphi} = H$ and for each neighborhood U of e there exists a syndetic subgroup A of G such that $A\varphi \subset U$. Then H is zero-dimensional.*

PROOF. Let U be a closed neighborhood of e. There exists a syndetic subgroup A of G such that $A\varphi \subset U$. Define $K = \overline{A\varphi}$. Now K is a subgroup of finite index in H. Thus K is an open-closed neighborhood of e such that $K \subset U$. The proof is completed.

5.29. THEOREM. *Let X be compact, let T be discrete, let Φ be the total homeomorphism group of X and let Φ be provided with its space-index topology. Then the following statements are equivalent:*

(1) *(X, T, π) is regularly almost periodic.*

(2) *The closure in Φ of the transition group $[\pi^t : X \to X \mid t \in T]$ is a zero-dimensional compact topological group.*

PROOF. Use **4.45**, 5.27, 5.28.

5.30. REMARK. Let φ be a continuous homomorphism of a topological group T into a zero-dimensional compact topological group S and let (S, T, ρ) be the {left}{right} transformation group of S induced by T under φ. Then:

(1) (S, T, ρ) is regularly almost periodic.

(2) S is a regularly almost periodic minimal orbit-closure under (S, T, ρ) if and only if $\overline{T\varphi} = S$.

5.31. THEOREM. *Let $x \in X$ and let T be discrete abelian. Then the following statements are pairwise equivalent.*

(1) *X is a regularly almost periodic compact minimal orbit-closure under (X, T, π).*

(2) *X is an almost periodic zero-dimensional compact minimal orbit-closure under (X, T, π).*

(3) *There exists a unique group structure of X which makes X a topological group such that (X, π_x) is a zero-dimensional compactification of T.*

(4–5) *There exist a zero-dimensional compact topological group S and a homomorphism $\varphi : T \to S$ such that $\overline{T\varphi} = S$ and the {left}{right} transformation group of S induced by T under φ is isomorphic to (X, T, π).*

PROOF. Use **4.48, 5.08** and **5.30.**

5.32. LEMMA. *Let G be a compact metrizable abelian group and let H be the set of all $x \in G$ such that the closure of $[x^n \mid n \in \mathcal{I}]$ is zero-dimensional. Then H is a dense subgroup of G.*

This result is from the theory of Lie groups.

5.33. THEOREM. *Let X be a compact metric space with metric ρ, let φ be an almost periodic homeomorphism of X onto X and let ϵ be a positive number. Then there exists a regularly almost periodic homeomorphism ψ of X onto X such that $x \in X$ implies $\rho(x\varphi, x\psi) < \epsilon$; indeed, ψ may be chosen to be the uniform limit of a sequence of {positive}{negative} powers of φ.*

PROOF. Use 5.32.

5.34. THEOREM. *Let φ be a regularly almost periodic homeomorphism of a two-dimensional manifold X onto X. Then φ is periodic.*

PROOF. Let $\mathfrak{A}$ be the partition space of all orbit-closures. By 5.08 the projection $\psi : X \to \mathfrak{A}$ is light interior. It is known (Whyburn [1], p. 191) that a light interior mapping on a two-dimensional manifold is locally finite-to-one. Hence φ is pointwise periodic. It is known (Montgomery [2]) that a pointwise periodic homeomorphism on a manifold is periodic. The proof is completed.

5.35. REMARK. The only universally valid implications among the regular almost periodicity properties are the obvious ones. These implications are summarized in Table 3.

TABLE 3

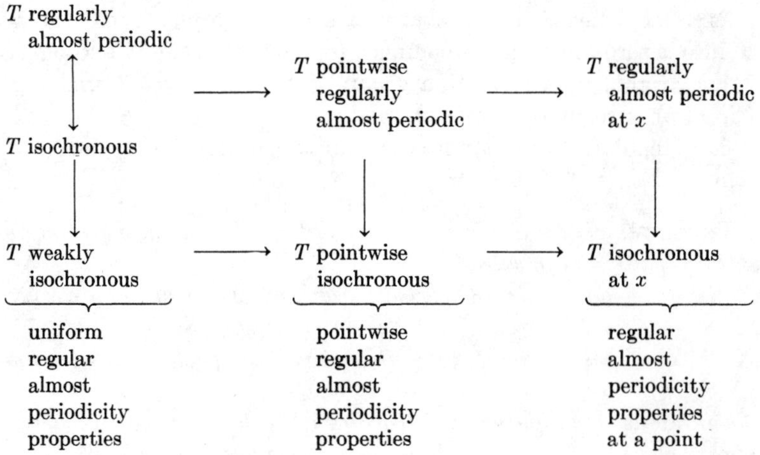

5.36. NOTES AND REFERENCES.

(5.04) Since T is discrete, this is a purely group-theoretic result and (2) may be found in Kurosch [1].

(5.08) Cf. P. A. Smith [1].

(5.09) This theorem and its proof are due to E. E. Floyd. (Personal communication.)

(5.18) Subject to the additional hypotheses that T is discrete and abelian, this theorem was proved by Garcia and Hedlund [1]. A proof of 5.18 was obtained by S. Schwartzman [1].

(5.33) Cf. P. A. Smith [1].

6. REPLETE SEMIGROUPS

6.01. Definition. A topological group T is said to be *generative* provided that T is abelian and is generated by some compact neighborhood of the identity element of T.

6.02. Standing notation. Throughout this section T denotes a generative topological group.

6.03. Remark. The assumption that T is generative ensures the existence of "sufficiently many" replete semigroups in T. The generality of generative topological groups T is indicated by the structure theorem [Weil [1], p. 110] that T is isomorphic to $C \times \mathcal{I}^n \times \mathcal{R}^m$ where C is a compact abelian group and n, m are nonnegative integers. The theorems of this section have lemma value. Some of them do not use all of the hypothesis that T is generative.

6.04. Theorem. *Let P be a semigroup in T. Then:*
(1) *If P is replete in T, then $T = P^{-1}P$.*
(2) *If $T = P^{-1}P$ and if T is discrete, then P is replete in T.*

Proof. (1) If $t \in T$, then there exists $s \in T$ such that st, $st^2 \in P$ whence $t = (st)^{-1}(st^2) \in P^{-1}P$.

(2) Let n be a positive integer and let $t_1, \cdots, t_n \in T$. For each $i = 1, \cdots, n$ choose p_i, $q_i \in P$ such that $t_i = p_i^{-1}q_i$. Define $t = p_1 \cdots p_n$. Then $tt_i \in P$ $(i = 1, \cdots, n)$. The proof is completed.

6.05. Remark. Let P be a semigroup in $\mathcal{R}$ such that P is maximal with respect to the property of containing only positive nonintegral numbers. Then P is replete relative to the discrete topology of $\mathcal{R}$ but P is not replete relative to the natural topology of $\mathcal{R}$.

6.06. Theorem. *Let P and Q be replete semigroups in T. Then PQ and P^{-1} are replete semigroups in T.*

Proof. Obvious.

6.07. Theorem. *Let P be a replete semigroup in T and let K be a compact subset of T such that $e \in K$. Then $\bigcap_{k \in K} kP$ is a replete semigroup in T.*

Proof. Since $Q = \bigcap_{k \in K} kP = \bigcap_{k \in K} (P \cap kP)$ and each $P \cap kP$ $(k \in K)$ is a semigroup, it follows that Q is a semigroup. Let C be a compact subset of T and define $D = C \cup K^{-1}C$. There exists $t \in T$ such that $tD \subset P$. Now $k \in K$ implies $C \subset D \cap kD$ whence $tC \subset tD \cap kt\,D \subset P \cap kP$. Thus $tC \subset Q$ and the proof is completed.

6.08. Theorem. *Let E be a subset of T such that E contains an open symmetric neighborhood of e which generates T and let $t \in T$. Then $\bigcup_{i=1}^{+\infty} t^iE^i$ is a replete semigroup in T.*

PROOF. Obvious.

6.09. THEOREM. *The class of all replete semigroups in T has a countable base, that is, there exists a sequence $P_1, P_2, \cdots$ of replete semigroups in T such that each replete semigroup in T contains P_n for some positive integer n.*

PROOF. By 6.03 there exist a compact subgroup K of T and a separable closed subgroup S of T such that $T = KS$. Let $[s_1, s_2, \cdots]$ be a countable dense subset of S and let U be a compact symmetric neighborhood of e whose interior generates T. Define $P_n = \bigcup_{i=1}^{+\infty} s_n^i U^i$ $(n = 1, 2, \cdots)$. Each P_n $(n = 1, 2, \cdots)$ is a replete semigroup in T. Let P be a replete semigroup in T. There exist a neighborhood V of e and $t \in T$ such that $VtKU \subset P$. Choose $k \in K$ and $s \in S$ for which $t = ks$. There exists a positive integer n such that $s_n \in Vs$. Hence $s_nU \subset VsU = Vtk^{-1}U \subset P$ and $P_n \subset P$. The proof is completed.

6.10. THEOREM. *Let S be a closed syndetic subgroup of T and let K be a compact subset of T. Then there exists a compact subset H of S such that $K^n \cap S \subset H^n$ for all integers n.*

PROOF. We may suppose that $T = SK$, $e \in K$ and $K = K^{-1}$. Define $H = K^3 \cap S$. Now H is a compact subset of S and to prove the theorem it is enough to show that $K^n \cap S \subset H^n$, for all positive integers n. Let n be a positive integer and let $k_1, \cdots, k_n \in K$ such that $k_1 \cdots k_n \in S$. If $t \in T$, then there exists $s \in S$ such that $t \in sK^{-1}$ whence $s \in tK$. Thus for each integer i $(1 \leqq i < n)$ there exists $s_i \in S$ such that $s_i \in k_1 \cdots k_iK$. Define $s_n = k_1 \cdots k_n$. Clearly $s_n \in S$ and $s_n \in k_1 \cdots k_nK$. Now $s_i^{-1}s_{i+1} \in Kk_{i+1}K \subset K^3$ and $s_i^{-1}s_{i+1} \in S$ $(1 \leqq i < n)$ whence $s_i^{-1}s_{i+1} \in H$ and $s_{i+1} \in s_iH$ $(1 \leqq i < n)$. Also $s_1 \in k_1K \subset K^3$ and $s_1 \in S$ whence $s_1 \in H$. We conclude that $k_1 \cdots k_n = s_n \in s_{n-1}H \subset s_{n-2}H^2 \subset \cdots \subset s_1H^{n-1} \subset H^n$. The proof is completed.

6.11. THEOREM. *Let S be a closed syndetic subgroup of T. Then S is a generative topological group.*

PROOF. Use 6.10.

6.12. REMARK. Actually 6.11 remains true when "syndetic" is omitted.

6.13. THEOREM. *Let S be a closed syndetic subgroup of T and let Q be a replete semigroup in S. Then there exists a replete semigroup P in T such that $P \cap S \subset Q$.*

PROOF. Let K be a compact symmetric neighborhood of e whose interior generates T. By 6.10 there exists a compact subset H of S such that $K^n \cap S \subset H^n$ for all integers n. For some $s \in S$ we have $sH \subset Q$. Define $P = \bigcup_{n=1}^{+\infty} s^nK^n$. Now P is a replete semigroup in T and

$$P \cap S \subset \bigcup_{n=1}^{+\infty} (s^nK^n \cap S) \subset \bigcup_{n=1}^{+\infty} s^nH^n \subset Q.$$

The proof is completed.

6.14. Theorem. *Let S be a syndetic subgroup of T and let P be a replete semigroup in T. Then $P \cap S$ is a replete semigroup in S.*

Proof. Clearly $P \cap S$ is a semigroup. Let H be a compact subset of S and let K be a compact subset of T for which $T = SK$. Since $K^{-1}H$ is compact, there exists $t \in T$ such that $tK^{-1}H \subset P$. Choose $s \in S$ and $k \in K$ such that $t = sk$. Now $sH = tk^{-1}H \subset P$ and $sH \subset S$. Hence $sH \subset P \cap S$ and the proof is completed.

6.15. Theorem. *Let $A \subset T$. Then the following statements are pairwise equivalent.*

(1) *A is extensive in T, that is, A intersects every replete semigroup in T.*
(2) *A intersects every translate of every replete semigroup in T.*
(3) *$T = AP$ for every replete semigroup P in T.*
(4) *$T = AtP$ for every replete semigroup P in T and every $t \in T$.*

Proof. Assume (1). We prove (2). Let P be a replete semigroup in T and let $t \in T$. By 6.07, $P \cap tP$ is a replete semigroup in T. Hence A intersects $P \cap tP$ and consequently tP.

Assume (2). We prove (3). Let P be a replete semigroup in T and let $t \in T$. Since $A \cap tP^{-1} \neq \emptyset$, there exist $a \in A$ and $p \in P$ such that $a = tp^{-1}$. Thus $t = ap \in AP$.

Clearly, (3) implies (4).

Assume (4). We prove (1). Let P be a replete semigroup in T. Since $T = AP^{-1}$, there exist $a \in A$ and $p \in P$ such that $e = ap^{-1}$. Hence $a = p$ and $A \cap P \neq \emptyset$. The proof is completed.

6.16. Theorem. *Let A be an extensive subset of T and let $t \in T$. Then tA and A^{-1} are extensive in T.*

Proof. Use 6.15 and 6.06.

6.17. Theorem. *Let A be a syndetic subset of T. Then A is extensive in T.*

Proof. Let P be a replete semigroup in T. Choose a compact subset K of T such that $T = AK$ and then choose $t \in T$ such that $tK \subset P$. Since $T = AtK \subset AP$, the conclusion follows from 6.15.

6.18. Theorem. *Let $A, B, K \subset T$ such that A is extensive in T, K is compact, and $A \subset BK$. Then B is extensive in T.*

Proof. Let P be a replete semigroup in T. Choose $t \in T$ such that $tK \subset P$. From 6.15 we conclude that $T = AtP \subset BtKP \subset BP$ and that B is extensive in T.

6.19. Theorem. *Let S be a closed syndetic subgroup of T and let $A \subset S$. Then A is extensive in S if and only if A is extensive in T.*

Proof. Use 6.13 and 6.14.

6.20. THEOREM. *Let S be a closed syndetic subgroup of T, let Q be a replete semigroup in S and let K be a compact subset of T such that $e \in K$. Then there exists a replete semigroup P in T such that $PK \cap S \subset Q$.*

PROOF. By 6.13 there exists a replete semigroup R in T such that $R \cap S \subset Q$. Define $P = \bigcap_{k \in K} Rk^{-1}$. By 6.07, P is a replete semigroup in T. Now $PK \subset R$ whence $PK \cap S \subset R \cap S \subset Q$. The proof is completed.

6.21. THEOREM. *Let S be a closed syndetic subgroup of T, let U be a neighborhood of e and let $t \in T$. Then there exists a positive integer n such that $t^n \in SU$.*

PROOF. Let K be a compact subset of T for which $T = SK$. Choose a neighborhood V of e such that $VV^{-1} \subset U$. Let $\mathfrak{F}$ be a finite class of translates of V which covers K. To each positive integer n there correspond $s_n \in S$ and $k_n \in K$ such that $t^n = s_n k_n$. Select positive integers p, q such that $p > q$ and $k_p, k_q \in V_0$ for some $V_0 \in \mathfrak{F}$. Then $t^{p-q} = s_p s_q^{-1} k_p k_q^{-1} \in Sk_p k_q^{-1} \subset SV_0 V_0^{-1} \subset SVV^{-1} \subset SU$. The proof is completed.

6.22. THEOREM. *Let S be a closed syndetic subgroup of T, let U be a neighborhood of e, let K be a compact subset of T and let $k_1, k_2, \cdots$ be a sequence of elements of K. Then there exist finitely many positive integers $i_1, \cdots, i_n$ $(n \geqq 1)$ such that $i_1 < \cdots < i_n$ and $k_{i_1} \cdots k_{i_n} \in SU$.*

PROOF. It follows readily from 6.21 that there exist finitely many open subsets $U_1, \cdots, U_m$ of T and positive integers $p_1, \cdots, p_m$ such that $K \subset \bigcup_{j=1}^m U_j$ and $U_j^{p_j} \subset SU$ $(j = 1, \cdots, m)$. There exists an integer j $(1 \leqq j \leqq m)$ such that $k_i \in U_j$ for infinitely many positive integers i. Define $n = p_j$. Choose positive integers $i_1, \cdots, i_n$ such that $i_1 < \cdots < i_n$ and $k_{i_1}, \cdots, k_{i_n} \in U_j$. Hence $k_{i_1} \cdots k_{i_n} \in U_j^n \subset SU$. The proof is completed.

6.23. STANDING NOTATION. For the remainder of this section (X, T, π) denotes a transformation group.

6.24. THEOREM. *Let Y be a subset of X such that every replete semigroup in T contains a compact set E such that $Y \subset YE$. Then there exists a compact subset C of T such that $YT = YC$.*

PROOF. Let U be an open symmetric neighborhood of e such that U generates T and $\overline{U}$ is compact. Define $H = \overline{U}^2$ and $K = \overline{U}^3 = \overline{U}H$.

We first show that there exists a positive integer n such that if $k \in K$, then $Y \subset \bigcup_{i=1}^n Yk(kH)^i$. To show this it is enough to prove that if $k_0 \in K$, then there exists a positive integer m and a neighborhood V of e such that $k \in k_0 V$ implies $Y \subset \bigcup_{i=1}^m Yk(kH)^i$. Now suppose $k_0 \in K$. Define $P = \bigcup_{i=1}^{+\infty} k_0(k_0 U)^i$. The set P is an open replete semigroup in T. Hence P contains a compact set E such that $Y \subset YE$. Choose a compact symmetric neighborhood V of e for which $V \subset U$ and $EV \subset P$. Since EV is compact, there exists a positive integer m such that $EV \subset \bigcup_{i=1}^m k_0(k_0 U)^i$ and hence $YV \subset YEV \subset \bigcup_{i=1}^m Yk_0(i_0 U)^i$.

Let $k \in k_0V$ and $y \in Y$. Choose $v \in V$ such that $k_0 = kv$. Then

$$yv \in \bigcup_{i=1}^{m} Yk_0(k_0U)^i = \bigcup_{i=1}^{m} Ykv(kvU)^i \subset \bigcup_{i=1}^{m} Ykv(kH)^i$$

and $y \in \bigcup_{i=1}^{m} Yk(kH)^i$. This completes the proof that there exists a positive integer n such that if $k \in K$, then $Y \subset \bigcup_{i=1}^{n} Yk(kH)^i$. Let n denote such an integer.

Choose a positive integer p ($p \geqq n$) so large that if $k_1, \cdots, k_{p+1} \in K$, then for some $n + 1$ of the elements $k_1, \cdots, k_{p+1}$, let us say $k_1, \cdots, k_{n+1}$, we have $k_i^{-1}k_j \in U$ ($i, j = 1, \cdots, n + 1$). We now show that $YT \subset YK^p$, which will complete the proof. Assume $YT \not\subset YK^p$. Then $YK^{p+1} \not\subset YK^p$ for otherwise $YT \subset \bigcup_{i=1}^{+\infty} YK^i \subset \bigcup_{i=p}^{+\infty} YK^i \subset YK^p$. Select $y \in Y$ and $k_1, \cdots, k_{p+1} \in K$ for which $yk_1 \cdots k_{p+1} \notin YK^p$. There exist $n + 1$ of the elements $k_1, \cdots, k_{p+1}$, let us say $k_1, \cdots, k_{n+1}$, such that $k_i^{-1}k_j \in U$ ($i, j = 1, \cdots, n + 1$). Let r be a positive integer such that $r \leqq n$. It follows that $yk_1 \cdots k_{r+1} \notin YK^r$ and $yk_1(k_1u_2) \cdots (k_1u_{r+1}) \notin YK^r$ where $u_2, \cdots, u_{r+1}$ are elements of U for which $k_2 = k_1u_2, \cdots, k_{r+1} = k_1u_{r+1}$. Thus $yk_1^{r+1} \notin YH^r$ and $y \notin Yk_1^{-1}(k_1^{-1}H)^r$. We conclude that $y \notin \bigcup_{i=1}^{n} Yk_1^{-1}(k_1^{-1}H)^i$. Since $k_1^{-1} \in K$, this contradicts the definition of n. The proof is completed.

6.25. Theorem. *Let U be an open subset of X such that $\overline{U}$ is compact and suppose that for every compact subset K of T there exist $x \in \overline{U}$ and $t \in T$ such that $xtK \cap U = \emptyset$. Then there exist $y \in \overline{U}$ and a replete semigroup P in T such that $yP \cap U = \emptyset$.*

Proof. Assume the conclusion is false. Then for each $x \in \overline{U}$ and each replete semigroup P in T we have $xP \cap U \neq \emptyset$ whence $x \in UP^{-1}$. Since the inverse of a replete semigroup in T is also a replete semigroup in T, it follows that for each replete semigroup P in T we have $\overline{U} \subset UP$; since U is open and $\overline{U}$ is compact, we can choose a finite subset F of P such that $\overline{U} \subset UF$. Thus each replete semigroup in T contains a finite set F for which $U \subset UF$. By 6.24 there exists a compact subset C of T such that $UT = UC$. Hence $x \in U$ and $t \in T$ implies $xt \in UC$ and $xtC^{-1} \cap U \neq \emptyset$. This contradicts the hypothesis. The proof is completed.

6.26. Theorem. *Let A be a nonvacuous subset of T such that every replete semigroup in T contains a compact set E such that $A \subset AE$. Then A is syndetic in T.*

Proof. Apply 6.24 to the transformation group (T, T, ρ) where $\rho : T \times T \to T$ is defined by $(t, s)\rho = ts$ ($t, s \in T$).

6.27. Theorem. *Let P be an extensive semigroup in T. Then P is syndetic in T.*

Proof. Let Q be a replete semigroup in T. Choose $t \in P \cap Q^{-1}$. Then $t^{-1} \in Q$ and $P \subset Pt^{-1}$. The conclusion now follows from 6.26.

6.28. THEOREM. *Let M be a closed subset of X, let P be a replete semigroup in T, let $MP \subset M$ and let $x \in M - \overline{MP}$. Then there exists a replete semigroup Q in T such that $Q \subset P$ and $x \notin \overline{xQ}$.*

PROOF. Let K be a compact subset of T such that K contains an open symmetric neighborhood of e which generates T. Select $t \in T$ such that $H = tK \subset P$. Define $Q = \bigcup_{n=1}^{+\infty} H^n$. Now $MH \subset MP \subset M$ whence $MH^n \subset MH$ $(n = 1, 2, \cdots)$ and $MQ \subset MH$. Since M is closed and H is compact, MH is closed. Then $\overline{xQ} \subset \overline{MQ} \subset MH \subset \overline{MP}$ and $x \notin \overline{xQ}$. Since Q is a replete semigroup in T, the proof is completed.

6.29. THEOREM. *Let S be a subgroup of T which is not syndetic in T, and let K be a compact subset of T. Then there exists a compact subset C of T such that every translate of C contains a translate of K disjoint from S.*

PROOF. We may suppose that K contains an open symmetric neighborhood of e which generates T. Assume the conclusion is false. Then for each positive integer n there exists $t_n \in T$ such that S intersects every translate of K which is contained in t_nK^n. It follows that there exists $s_n \in t_nK \cap S$ $(n = 1, 2, \cdots)$. Since S is a group, every translate of K contained in $s_n^{-1}t_nK^n$ for some positive integer n intersects S. We show S is syndetic in T. Let K_0 be an arbitrary translate of K. Choose a positive integer m such that $K_0K \subset K^m$. Since $s_mt_m^{-1} \in K$, we have $K_0s_mt_m^{-1} \subset K^m$ and $K_0 \subset s_m^{-1}t_mK^m$. It follows that $K_0 \cap S \neq 0$. Thus S is syndetic in T, contrary to hypothesis.

6.30. THEOREM. *Let n be a positive integer, let $S_1, \cdots, S_n$ be subgroups of T such that each is not syndetic in T, let P be a replete semigroup in T and let K_0 be a compact subset of T. Then there exists a translate of K_0 contained in P and disjoint from $\bigcup_{i=1}^{n} S_i$.*

PROOF. By 6.29 there exist compact subsets $K_1, \cdots, K_n$ of T such that for each $i = 1, \cdots, n$ every translate of K_i contains a translate of K_{i-1} disjoint from S_i. Since P contains some translate of K_n, the conclusion follows.

6.31. THEOREM. *Suppose T is not compact. Then there exists a replete semigroup P in T such that $e \notin P$.*

PROOF. For otherwise $[e]$ would be syndetic in T by 6.26 and thus T would be compact.

6.32. THEOREM. *Suppose T is not compact, let P be a replete semigroup in T and let K be a compact subset of T. Then there exists a replete semigroup Q in T such that $Q \subset P$ and $Q \cap K = \emptyset$.*

PROOF. By 6.31 we may suppose that $P \neq T$. It is enough to show that $pP \cap K = \emptyset$ for some $p \in P$ and then to take $Q = pP$. Assume that $p \in P$ implies $pP \cap K \neq \emptyset$ whence $p \in KP^{-1}$. Thus $P \subset KP^{-1}$, $T = PP^{-1} \subset KP^{-1}P^{-1} \subset KP^{-1}$ and $T = PK^{-1}$. Choose $t \in T$ such that $K^{-1}t \subset P$; we have $T = Tt = PK^{-1}t \subset PP \subset P$. This is a contradiction. The proof is completed.

6.33. Definition. Let $x \in X$ and let $P \subset T$. The *P-limit set of x*, denoted P_x, is defined to be $\bigcap_{t \in T} \overline{xtP}$. Each point of P_x is called a *P-limit point of x*.

6.34. Remark. Let $x \in X$, and let P be a replete semigroup in T. Then:
(1) If $t \in T$, then $P_x = P_{xt} = P_x t$.
(2) P_x is closed invariant.
(3) $P_x = \bigcap_{p \in P} \overline{xpP} \subset \overline{xP}$.
(4) If X is compact, then $P_x \neq \emptyset$.
(5) If T is connected, then $xT \cup P_x$ is connected.
(6) If P is connected, then $xP \cup P_x$ is connected.
(7) If X is compact and if P is connected, then P_x is connected.

6.35. Definition. Let S be a replete semigroup in T and let $A \subset T$. The set A is said to be *S-extensive* provided that if P is a replete semigroup in T such that $P \subset S$, then $A \cap P \neq \emptyset$. The set A is said to be *S-syndetic* provided there exists a compact subset K of S such that $S \subset AK$.

6.36. Definition. Let S be a replete semigroup in T. If in 3.13 the term *admissible* is replaced by {*S-extensive*} {*S-syndetic*}, then the term *recursive* is replaced by {*S-recurrent*} {*S-almost periodic*}.

6.37. Notes and references.

A number of the theorems of this chapter can be found in Gottschalk and Hedlund [10].

7. RECURRENCE

7.01. Standing notation. Throughout this section (X, T, π) denotes a transformation group whose phase group T is generative.

7.02. Remark. Let $x \in X$. Then the following statements are pairwise equivalent:

(1) T is recurrent at x; that is to say, if U is a neighborhood of x, then there exists an extensive subset A of T such that $xA \subset U$.

(2) If U is a neighborhood of x and if P is a replete semigroup in T, then $xP \cap U \neq \emptyset$.

(3) If U is a neighborhood of x, if P is a replete semigroup in T and if $t \in T$, then $xtP \cap U \neq \emptyset$.

(4) If U is a neighborhood of x and if P is a replete semigroup in T, then $xT \subset UP$.

(5) If P is a replete semigroup in T, then $x \in \overline{xP}$.

(6) If P is a replete semigroup in T and if $t \in T$, then $x \in \overline{xtP}$.

(7) If P is a replete semigroup in T, then $\overline{xT} = \overline{xP}$.

7.03. Theorem. *Let $x \in X$ and let T be recurrent at x. Then T is recurrent at every point of $\overline{xT}$.*

Proof. Use 3.21.

7.04. Inheritance theorem. *Let S be a closed syndetic subgroup of T. Then:*

(1) If $x \in X$, then S is recurrent at x if and only if T is recurrent at x.

(2) S is pointwise recurrent if and only if T is pointwise recurrent.

Proof. Use 3.36.

7.05. Theorem. *Let X be locally compact, let T be pointwise recurrent and let the class of all orbit-closures under T be a partition of X. Then T is pointwise almost periodic.*

Proof. We may suppose that X is minimal under T. Assume some point x of X is not almost periodic. Then there exists an open neighborhood U of x such that $\overline{U}$ is compact and such that for each compact subset K of T there exists $t \in T$ for which $xtK \cap U = \emptyset$. By 6.25 there exists $y \in \overline{U}$ and a replete semigroup P in T such that $yP \cap U = \emptyset$. Hence $x \notin \overline{yP}$. Since y is recurrent, $\overline{yP} = \overline{yT}$. Therefore $x \notin \overline{yT}$. This contradicts the minimality of X. The proof is completed.

7.06. Theorem. *Let $T = \mathcal{I}$ or $\mathcal{R}$, let S be a locally compact topological group and let φ be a continuous homomorphism of T into S. Then exactly one of the following statements is valid:*

(1) φ is a homeomorphic isomorphism of T into S.

(2) $\overline{T\varphi}$ *is compact and for each neighborhood* V *of the identity of* S *the set* $V\varphi^{-1}$ *is syndetic in* T.

Proof. Apply 7.05 and 4.09 to the {left}{right} transformation group of S induced by T under φ.

7.07. Theorem. *Let* X *be locally compact zero-dimensional and let* T *be pointwise recurrent. Then* T *is locally weakly almost periodic.*

Proof. Let $x \in X$ and let U be a neighborhood of x. Choose an open compact neighborhood V of x such that $V \subset U$. It is enough to show that there exists a compact subset K of T such that $y \in V$ and $t \in T$ implies $ytK \cap U \neq \emptyset$. Assume this to be false. Then by 6.25 there exists $y \in V$ and a replete semigroup P in T such that $yS \cap V = \emptyset$. Hence y is not recurrent. This is a contradiction and the proof is completed.

7.08. Theorem. *Let* X *be a compact zero-dimensional uniform space and let* T *be pointwise recurrent. Then* T *is weakly almost periodic.*

Proof. Use 7.07 and 4.23(2).

7.09. Remark. Let $x \in X$ and let T be almost periodic at x. Then T is recurrent at x.

Proof. Use 6.17.

7.10. Theorem. *Let* X *be a compact zero-dimensional uniform space, let* T *be discrete and let* T *be pointwise regularly almost periodic. Then* T *is regularly almost periodic.*

Proof. Use 5.17, 7.04 and 7.08.

7.11. Theorem. *Let* X *be a compact metric space and let* T *be discrete. Then* T *is regularly almost periodic if and only if* T *is pointwise regularly almost periodic and weakly almost periodic.*

Proof. The necessity is trivial. We prove the sufficiency. Let S be a syndetic subgroup of T. By 4.24 and 5.17 it is enough to show that the class $\mathfrak{A}$ of all orbit-closures under S is a star-closed partition of X. By 4.08 and 5.07, $\mathfrak{A}$ is a partition of X. We show $\mathfrak{A}$ is star-closed. Let $x_0, x_1, x_2, \cdots$ be a sequence of points of X such that $\lim_{n\to\infty} x_n = x_0$. Since $\lim_{n\to\infty} \overline{x_nT} = \overline{x_0T}$ by 1.38 and 4.24, $Y = \bigcup_{m=0}^{+\infty} \overline{x_mT}$ is closed in X. By 5.08, each set $\overline{x_mT}$ $(m = 0, 1, \cdots)$ is zero-dimensional and therefore Y is a T-invariant compact zero-dimensional subset of X. It follows from 5.07 that S is pointwise regularly almost periodic on Y and it then follows from 7.08 that S is weakly almost periodic on Y. Hence $\lim_{n\to\infty} \overline{x_nS} = \overline{x_0S}$. The proof is completed.

7.12. Remark. Let $x \in X$. Then the following statements are pairwise equivalent:

(1) T is regionally recurrent at x; that is to say, if U is a neighborhood of x, then there exists an extensive subset A of T such that $a \in A$ implies $U \cap Ua \neq \emptyset$.

(2–3) If U is a neighborhood of x and if P is a replete semigroup in T, then $\{U \cap UP \neq \emptyset\}\{x \in \overline{UP}\}$.

(4–5) If U is a neighborhood of x, if P is a replete semigroup in T and if $t \in T$, then $\{U \cap UtP \neq \emptyset\}\{x \in \overline{UtP}\}$.

(6) If U is a neighborhood of x and if P is a replete semigroup in T, then $xT \subset \overline{UP}$.

7.13. Theorem. *Let R be the set of all regionally T-recurrent points of X. Then R is a T-invariant closed subset of X.*

Proof. Use 3.22 and 3.28.

7.14. Theorem. *Let $x \in X$ and let T be regionally recurrent at x. Then T is regionally recurrent at every point of $\overline{xT}$.*

Proof. Use 7.13.

7.15. Theorem. *Let T be regionally recurrent and let R be the set of all T-recurrent points of X. Then:*

(1) *If X is metrizable, then R is a T-invariant residual G_δ subset of X.*

(2) *If X is a complete metric space, then $X = \overline{R}$.*

Proof. Use 3.31 and 6.09.

7.16. Theorem. *Let X be a complete metric space. Then T is regionally recurrent if and only if the set of all T-recurrent points of X is a dense subset of X.*

Proof. Use 3.28 and 7.15.

7.17. Definition. The *center* of the transformation group (X, T) is defined to be the greatest T-invariant subset C of X such that T is regionally recurrent on C, that is, such that the transformation group (C, T) is regionally recurrent; the existence of C is readily seen, for C is the union of the class of all T-invariant subsets of X on which T is regionally recurrent.

7.18. Theorem. *Let C be the center of (X, T). Then:*

(1) *C is a closed subset of X.*

(2) *C is contained in the set of all T-regionally recurrent points of X.*

Proof. Obvious.

7.19. Remark. For a T-invariant subset Y of X let Y^* denote the set of all points of Y at which the transformation group (Y, T) is regionally recurrent. For each ordinal α define R_α by transfinite induction as follows:

(I) $R_0 = X$.

(II) If α is a successor ordinal, then $R_\alpha = R^*_{\alpha-1}$.

(III) If α is a limit ordinal, then $R_\alpha = \bigcap_{\beta<\alpha} R_\beta$.

Then:
(1) There exists a least ordinal γ such that $R_\gamma = R_\alpha$ for all $\alpha > \gamma$.
(2) If X is second-countable, then γ is countable.
(3) R_γ is the center of (X, T).

7.20. THEOREM. *Let X be a complete metric space. Then the center of (X, T) coincides with the closure of the set of all T-recurrent points of X.*

PROOF. Use 3.28 and 7.16.

7.21. INHERITANCE THEOREM. *Let S be a closed syndetic subgroup of T. Then S is regionally recurrent if and only if T is regionally recurrent.*

PROOF. It follows from 6.19 that the regional recurrence of S implies the regional recurrence of T.

Now assume that T is regionally recurrent. Let $x \in X$, let U be an open neighborhood of x and let Q be a replete semigroup in S. There exist an open neighborhood U_0 of x and a compact symmetric neighborhood V of e in T such that $U_0V \subset U$. By 6.20 we can find a replete semigroup P in T such that $PV \cap S \subset Q$. For $i = 1, 2, \cdots$ we proceed inductively as follows:

There exists $p_i \in P$ such that $U_{i-1} \cap U_{i-1}p_i \neq \emptyset$. There exists $x_i \in U_{i-1}$ such that $x_ip_i \in U_{i-1}$. There exists an open neighborhood U_i of x_i such that $U_i \subset U_{i-1}$ and $U_ip_i \subset U_{i-1}$.

There exists a compact subset K of T such that $T = SK$. For each positive integer i there exist $s_i \in S$ and $k_i \in K$ such that $p_i = s_ik_i$. By 6.22 we can find positive integers $n, i_1, \cdots, i_n$ $(n \geqq 1)$ for which $i_1 < \cdots < i_n$ and $k_{i_1} \cdots k_{i_n} \in SV$. Since $U_{i_j}p_{i_j} \subset U_{i_{j-1}}$ $(j = 2, \cdots, n)$ and $U_{i_1}p_1 \subset U_0$, we conclude that $U_{i_n}p_{i_n} \cdots p_{i_1} \subset U_0$. Also $U_{i_n} \subset U_0$ whence $U_0 \cap U_0p_{i_n} \cdots p_{i_1} \neq \emptyset$. Now $p_{i_n} \cdots p_{i_1} = sk_{i_n} \cdots k_{i_1}$ where $s = s_{i_n} \cdots s_{i_1} \in S$. Choose $s_0 \in S$ and $v \in V$ such that $k_{i_n} \cdots k_{i_1} = s_0v$. Then $p_{i_n} \cdots p_{i_1} = ss_0v$. Define $q = p_{i_n} \cdots p_{i_1}v^{-1}$. We observe that $q = ss_0$, $q \in PV$, $ss_0 \in S$ and $PV \cap S \subset Q$. Hence $q \in Q$. Now $U_0v^{-1} \cap U_0p_{i_n} \cdots p_{i_1}v^{-1} \neq \emptyset$, $U_0 \subset U$ and $U_0v^{-1} \subset U$. Thus $U \cap Uq \neq \emptyset$ and the proof is completed.

7.22. DEFINITION. Let $x \in X$. The transformation group T is said to be *wandering at x* and the point x is said to be *wandering under T* or *T-wandering* provided that T is not regionally recurrent at x; that is, there exist a neighborhood U of x and a replete semigroup P in T such that $U \cap UP = \emptyset$.

7.23. REMARK. Let W be the set of all T-wandering points of X. Then W is a T-invariant open subset of X.

7.24. THEOREM. *Let X be a complete metric space and let $\{R\}\{W\}$ be the set of all $\{T$-recurrent$\}\{T$-wandering$\}$ points of X. Then $R \cup W$ is a residual subset of X.*

PROOF. Since T is regionally recurrent on $X - \overline{W}$, it follows from 7.20 that $\overline{R} \supset X - \overline{W}$, whence $X = \overline{R} \cup \overline{W}$ and $X - (R \cup W) \subset (\overline{R} - R) \cup$

$(\overline{W} - W)$. By 7.15, $\overline{R} - R$ is a first category subset of R and therefore of X. Since W is open in X, $\overline{W} - W$ is nowhere dense in X. The conclusion follows.

7.25. NOTES AND REFERENCES.

(7.02) If $T = \mathfrak{R}$, and thus (X, T, π) is a one-parameter continuous flow, the point $x \in X$ is recurrent if and only if, given any neighborhood U of x, there exists a sequence $\cdots < t_{-1} < t_0 < t_1 < \cdots$ in $\mathfrak{R}$ with $\lim_{n\to-\infty} t_n = -\infty$ and $\lim_{n\to+\infty} t_n = +\infty$ such that $xt_n \in U$ $(n \in \mathfrak{g})$. The classic term for *recurrent* is *stable in the sense of Poisson.*

(7.12) If $T = \mathfrak{R}$, the point $x \in X$ is regionally recurrent under T if and only if x is a non-wandering point in the sense of G. D. Birkhoff [2].

(7.15) This is the topological analogue of the Poincaré Recurrence Theorem (cf., eg., Carathéodory [1]). See also Hilmy [4].

(7.17–7.20) If $T = \mathfrak{R}$, the center coincides with Birkhoff's set of *central motions.* The process of 7.19, whereby the center is obtained by transfinite induction, is due to Birkhoff.

(7.21) Cf. Erdös and Stone [1].

8. INCOMPRESSIBILITY

8.01. STANDING NOTATION. Throughout this section (X, T, π) denotes a transformation group whose phase group T is generative.

8.02. THEOREM. *The following statements are pairwise equivalent:*

(1) *T is pointwise periodic.*

(2) *If M is a nonvacuous subset of X, then there exists an extensive subset A of T such that $a \in A$ implies $M \cap Ma \neq \emptyset$.*

(3) *If M is a subset of X, if P is a replete semigroup in T and if $MP \subset M$, then $MP = M$.*

PROOF. That (1) implies (2) follows from 6.17. Clearly (2) implies (3). Assume (3). We prove (1). Let $x \in X$ and let E be the period of x under T. By 6.27 it is enough to show that E is extensive in T. Let P be a replete semigroup in T. Define $M = x \cup xP$. Then $MP \subset M$ whence $MP = M$ and $xp = x$ for some $p \in P$. Therefore $E \cap P \neq \emptyset$. The proof is completed.

8.03. REMARK. Let $T = \mathcal{g}$ and let $0 \neq t \in T$. Then the following statement may be adjoined to 8.02.

(4) If $M \subset X$ such that $Mt \subset M$, then $Mt = M$.

8.04. LEMMA. *Let S be a semigroup in T and let $D = [x \mid x \in X, x \notin xS]$. Then there exists $C \subset X$ such that $C \cap CS = \emptyset$ and $D = CT$.*

PROOF. Let C be a maximal subset of X for which $C \cap CS = \emptyset$. It is easily verified that $D = CT$.

8.05. THEOREM. *The following statements are pairwise equivalent:*

(1) *The set of all T-periodic points of X is a residual subset of X.*

(2) *If M is a second category subset of X, then there exists an extensive subset A of T such that $a \in A$ implies $M \cap Ma \neq \emptyset$.*

(3) *If M is a subset of X, if P is a replete semigroup in T and if $MP \subset M$, then $M - MP$ is a first category subset of X.*

PROOF. Let E be the set of all T-periodic points of X.

Assume (1). We prove (3). Let $M \subset X$, let P be a replete semigroup in T, and let $MP \subset M$. Since $M - MP \subset X - E$, it follows that $M - MP$ is first category.

Assume (3). We prove (2). Let $M \subset X$, let P be a replete semigroup in T and let $M \cap MP = \emptyset$. It is enough to show that M is first category. If $N = M \cup MP$, then $NP \subset N$ whence $M = N - NP$ is first category.

Assume (2). We prove (1). By 6.09 there exists a countable base $P_1, P_2, \cdots$ for the replete semigroups in T. Let $D_i = [x \mid x \in X, x \notin xP_i]$ $(i = 1, 2, \cdots)$. Since $E = X - \bigcup_{i=1}^{+\infty} D_i$, it is enough to show that each $D_i (i = 1, 2, \cdots)$ is first category.

Let i be a positive integer. Write $P_i = P$ and $D_i = D$. By 8.04 there exists $C \subset X$ such that $C \cap CP = \emptyset$ and $D = CT$. Let V be a compact symmetric neighborhood of e. Define $Q = \bigcap_{t \in V^2} tP$. By 6.07, Q is a replete semigroup in T. Now $CV \cap CVQ = \emptyset$. If $M = CV \cup CVQ$, then $MQ \subset M$ whence $CV = M - MQ$ is first category. By 6.03 there exists a countable subset B of T such that $T = VB$. Hence $D = CT = CVB$ is first category. The proof is completed.

8.06. Remark. Let $T = \mathcal{I}$ and let $0 \neq t \in T$. Then the following statement may be adjoined to 8.05:

(4) If $M \subset X$ such that $Mt \subset M$, then $M - Mt$ is a first category subset of X.

8.07. Theorem. *The following statements are pairwise equivalent.*

(1) *T is pointwise recurrent.*

(2–3) *If M is {an open}{a closed} subset of X, if P is a replete semigroup in T and if $MP \subset M$, then $MP = M$.*

Proof. If $M \subset X$ and if $P \subset T$, then $\{MP \subset M\}\{MP = M\}$ if and only if $\{M'P^{-1} \subset M'\}\{M'P^{-1} = M'\}$. This shows the equivalence of (2) and (3). By 6.28, (1) implies (3). Assume (3). We prove (1). Let $x \in X$ and let P be a replete semigroup in T. Define $M = \overline{x \cup xP}$. Now $MP \subset \overline{xP} \subset M$ whence $MP = M$, $M = \overline{xP}$ and $x \in \overline{xP}$. The proof is completed.

8.08. Remark. Let $T = \mathcal{I}$ or $\mathcal{R}$ and let $0 \neq t \in T$. Then the following statement may be adjoined to 8.07:

(4) If M is an open or closed subset of X such that $Mt \subset M$, then $Mt = M$.

8.09. Theorem. *Let X be a complete metric space. Then the following statements are pairwise equivalent:*

(1) *The set of all T-recurrent points of X is a residual subset of X.*

(2–3) *If M is a second category {open}{closed} subset of X, then there exists an extensive subset A of T such that $a \in A$ implies $M \cap Ma \neq \emptyset$.*

(4–5) *If M is {an open}{a closed} subset of X, if P is a replete semigroup in T and if $MP \subset M$, then $M - MP$ is a first category subset of X.*

Proof. By 7.15 and 7.16, (1) is equivalent to regional recurrence of T. Clearly (2) and (3) are each equivalent to regional recurrence of T.

Assume (1). We prove (4). Let M be an open subset of X, let P be a replete semigroup in T and let $MP \subset M$. Suppose $M - MP$ is a second category subset of X. Then $M - MP$ is a second category subset of M. Since $M - MP$ is closed in M, there exists a nonvacuous open subset U of M, and therefore of X, such that $U \cap MP = \emptyset$. This contradicts (1).

By 6.28, (1) implies (5).

Assume (4). We prove T is regionally recurrent. Suppose T is not regionally recurrent. Then there exist a nonvacuous open subset U of X and a replete semigroup P in T such that $U \cap UP = \emptyset$. Define $M = U \cup UP$. Since M is open and $MP \subset M$, it follows that $U = M - MP$ is first category. This is a contradiction.

Assume (5). We prove T is regionally recurrent. Suppose T is not regionally recurrent. Then there exist a nonvacuous open subset U of X and a replete semigroup P in T such that $U \cap UP = \emptyset$ whence $U \cap \overline{UP} = \emptyset$. Define $M = \overline{U \cup UP}$. Since M is closed and $MP \subset \overline{UP} \subset M$, it follows that $M - MP$ is first category, $U \subset M - MP$ and U is first category. This is a contradiction. The proof is completed.

8.10. Remark. Let $T = \mathscr{I}$ or $\mathscr{R}$ and let $0 \neq t \in T$. Then the following statements may be adjoined to 8.09:

(6–7) If M is {an open}{a closed} subset of X such that $Mt \subset M$, then $M - Mt$ is a first category subset of X.

8.11. Definition. Let $x \in X$ and let $P \subset T$. The point x is said to be *compactive under* P or P-*compactive* provided that $\overline{xP}$ is compact. The antonym of "compactive" is "noncompactive."

8.12. Definition. Let $x \in X$. The point x is said to be *totally noncompactive* (*under* T) provided that x is P-noncompactive for every replete semigroup P in T.

8.13. Theorem. *Let X be a T_2-space, let R denote the set of all T-recurrent points of X, let N_1 denote the set of all totally noncompactive points of X and let N_2 denote the set of all T-noncompactive points of X. Consider the following statements*:

(I) $R \cup N_1 = X$.

(I′) $R \cup N_1$ *is a residual subset of* X.

(II) *If M is a compact subset of X, if P is a replete semigroup in T and if $MP \subset M$, then $MP = M$.*

(II′) *If M is a compact subset of X, if P is a replete semigroup in T and if $MP \subset M$, then $M - MP$ is a first category subset of X.*

(III) $R \cup N_2 = X$.

(III′) $R \cup N_2$ *is a residual subset of* X.

Then:

(1) I *implies* II; II *implies* III; I′ *implies* II′.

(2) *If X is a locally compact separable metrizable space, then* II′ *implies* III′.

Proof. (1) That I implies II, and II implies III follows from 6.28.

Assume I′. We prove II′. Let M be a compact subset of X and let P be a replete semigroup in T such that $MP \subset M$. If $x \in M - MP$, it follows from 6.28 that $x \notin R$ and it follows from $\overline{xP} \subset \overline{MP} \subset \overline{M} = M$ that $x \notin N_1$. Thus $M - MP \subset X - R \cup N_1$ and $M - MP$ is a first category subset of X. The proof that I′ implies II′ is completed.

(2) Let X be a locally compact separable metrizable space. Assume II′. We prove III′.

Let ρ be a compatible metric in X. Let $S_1, S_2, \cdots$ be a sequence of replete semigroups in T such that if S is any replete semigroup in T, then $S_n \subset S$ for some positive integer n. The existence of such a sequence is assured by 6.09. Let $X_1 \subset X_2 \subset \cdots$ be an expanding sequence of compact subsets of X such that if M is any compact subset of X, then $M \subset X_n$ for some positive integer n.

Let $\epsilon_1 > \epsilon_2 > \cdots$ be a decreasing sequence of positive numbers such that $\lim_{i\to\infty} \epsilon_i = 0$. For positive integers i, j, k let $E(i, j, k) = [x \mid x \in X, \overline{xT} \subset X_i, xS_j \cap x\epsilon_k = \emptyset]$. It is easily verified that $E(i, j, k)$ is a closed set and that $X - R \cup N_2 = \bigcup_{i,j,k=1}^{+\infty} E(i, j, k)$.

To prove III′ it is sufficient to prove that if i, j, k are positive integers, then $\text{int}\, E(i, j, k) = \emptyset$. Suppose that there exist positive integers i, j, k such that $\text{int}\, E(i, j, k) \neq \emptyset$. Then there exists an open subset Y of X such that $\emptyset \neq Y \subset E(i, j, k)$ and $\text{diam}\, Y < \epsilon_k$. Now $Y \cap YS_j = \emptyset$. For otherwise there would exist $s \in S_j$ and $y \in Y$ such that $ys \in Y$; but then $\rho(y, ys) < \epsilon_k$ and $yS_j \cap y\epsilon_k \neq \emptyset$, which is not the case. Since Y is open, $Y \cap \overline{YS_j} = \emptyset$. Let $M = \overline{Y \cup YS_j}$. Then $M \subset X_i$, M is a compact subset of X, $MS_j \subset \overline{YS_j} \subset M$ and $Y \subset M - MS_j$. It follows from II′ that Y is a set of the first category, which is not the case, since Y is nonvacuous open in X.

The proof is completed.

8.14. Remark. Let X be a locally compact separable metrizable space, let R be the set of all T-recurrent points of X, let N be the set of all totally noncompactive points of X and let $T = \mathcal{I}$ or $\mathcal{R}$. Then it may be proved that the following statements are pairwise equivalent:

(1) $R \cup N$ is a residual subset of X.

(2–3) If M is {an open}{a closed} subset of X such that $\overline{M}$ is compact, if $t \in T$ and if $Mt \subset M$, then $M - Mt$ is a first category subset of X.

8.15. Notes and references.

Many of the theorems of this section can be found in C. W. Williams [1]. See also Whyburn [1], Chapter XII.

(8–14) Cf. Hilmy [3] and E. Hopf [1].

9. TRANSITIVITY

9.01. STANDING NOTATION. Throughout this section (X, T, π) denotes a transformation group.

9.02. DEFINITION. Let $x \in X$. The transformation group (X, T) is said to be *transitive at* x and the point x is said to be *transitive under* (X, T) provided that if U is a nonvacuous open subset of X, then there exists $t \in T$ such that $xt \in U$.

The transformation group (X, T) is said to be {*pointwise*}{*point*} *transitive* provided that (X, T) is transitive at {every}{some} point of X.

The transformation group (X, T) is said to be (*regionally*) *transitive* provided that if U and V are nonvacuous open subsets of X, then there exists $t \in T$ such that $Ut \cap V \neq \emptyset$.

Let $x \in X$. The transformation group (X, T) is said be be *extensively transitive at* x and the point x is said to be *extensively transitive under* (X, T) provided that if U is a nonvacuous open subset of X, then there exists an extensive subset A of T such that $xA \subset U$.

The transformation group (X, T) is said to be *universally transitive* provided that if $x, y \in X$, then there exists $t \in T$ such that $xt = y$.

The transformation group (X, T) is said to be *extensively* (*regionally*) *transitive* provided that if U and V are nonvacuous open subsets of X, then there exists an extensive subset A of T such that $t \in A$ implies $Ut \cap V \neq \emptyset$.

The transformation group (X, T) is said to be {*pointwise*}{*point*} *extensively transitive* provided that (X, T) is extensively transitive at {every}{some} point of X.

The transformation group (X, T) is said to be (*regionally*) *mixing* provided that if U and V are nonvacuous open subsets of X, then there exists a compact subset K of T such that $t \in T - K$ implies $Ut \cap V \neq \emptyset$.

9.03. REMARK. The following statements are pairwise equivalent.

(1) (X, T) is universally transitive; that is to say, if $x, y \in X$, then there exists $t \in T$ such that $xt = y$.

(2) $xT = X$ for every $x \in X$.

(3) $xT = X$ for some $x \in X$.

9.04. REMARK. Let S be a subgroup of a topological group T. Then the {left}{right} transformation group of $\{T/S\}\{T\backslash S\}$ induced by T is universally transitive.

9.05. THEOREM. *Let (X, T, π) be universally transitive, let $x \in X$, let P be the period of x and let the motion $\pi_x : T \to X$ be open. Then (X, T, π) is isomorphic to the right transformation group of $T\backslash P$ induced by T.*

PROOF. Use 3.08.

9.06. REMARK. Let $x \in X$. Then the following statements are pairwise equivalent.

(1) (X, T) is transitive at x; that is to say, if U is a nonvacuous open subset of X, then there exists $t \in T$ such that $xt \in U$.

(2–3) If U is a nonvacuous open subset of X, then $\{xT \cap U \neq \emptyset\}\{x \in UT\}$.

(4) $\overline{xT} = X$.

9.07. REMARK. The transformation group (X, T) is pointwise transitive if and only if X is minimal under (X, T).

9.08. LEMMA. *Let* $(X, \mathfrak{U})$ *be a uniform space, let* $\mathfrak{V}$ *be a base of* $\mathfrak{U}$ *and let* A *be a dense subset of* X. *Then* $[\bigcup_{a \in A} a\beta \times a\beta \mid \beta \in \mathfrak{V}]$ *is a base of* $\mathfrak{U}$.

PROOF. Let $\beta \in \mathfrak{V}$. There exists $\gamma \in \mathfrak{V}$ such that $\gamma^{-1}\gamma \subset \beta$. Then $\bigcup_{a \in A} a\gamma \times a\gamma \subset \gamma^{-1}\gamma \subset \beta$. We show $\gamma \subset \bigcup_{a \in A} a\beta \times a\beta$. Let $(x_1, x_2) \in \gamma$. There exists $a \in A$ such that $a \in x_1\gamma$. Hence $(a, x_1) \in \gamma^{-1} \subset \beta$, $x_1 \in a\beta$, $(a, x_2) = (a, x_1) \cdot (x_1, x_2) \in \gamma^{-1}\gamma \subset \beta$, $x_2 \in a\beta$ and $(x_1, x_2) \in a\beta \times a\beta$. The proof is completed.

9.09. THEOREM. *Let* X *be a topological space and let* Φ *be a pointwise transitive homeomorphism group of* X. *Then the following statements are equivalent*:

(1) *There exists a compatible uniformity* $\mathfrak{U}$ *of* X *such that* Φ *is uniformly equicontinuous relative to* $\mathfrak{U}$.

(2) *For each* $x \in X$ *and each neighborhood* U *of* x *there exists a neighborhood* V *of* x *such that* $\varphi \in \Phi$ *with* $x \in V\varphi$ *implies* $V\varphi \subset U$.

If such a uniformity $\mathfrak{U}$ *exists, that is, if* (1) *holds, then* $[\bigcup_{\varphi \in \Phi} N\varphi \times N\varphi \mid N \in \mathfrak{N}]$ *is a base of* $\mathfrak{U}$, *where* $\mathfrak{N}$ *is any neighborhood base of any point of* X, *and* $\mathfrak{U}$ *is therefore unique.*

PROOF. For $A \subset X$ let $A^* = \bigcup_{\varphi \in \Phi} A\varphi \times A\varphi$. We remark that if $x \in X$ and if $A \subset X$, then $xA^* = \bigcup[A\varphi \mid \varphi \in \Phi, x \in A\varphi]$.

Assume that for each $x \in X$ and each neighborhood U of x there exists a neighborhood V of x such that $\varphi \in \Phi$ with $x \in V\varphi$ implies $V\varphi \subset U$. Let $x_0 \in X$, let $\mathfrak{N}$ be an open-neighborhood-base of x_0, and let $\mathfrak{U}$ be the filter on $X \times X$ generated by the filter-base $[N^* \mid N \in \mathfrak{N}]$. Let $U \in \mathfrak{N}$. Then there exists $V \in \mathfrak{N}$ such that $\varphi \in \Phi$ with $x_0 \in V\varphi$ implies $V\varphi \subset U$. To prove $\mathfrak{U}$ is a uniformity of X we need only show $V^*V^* \subset U^*$. Let $\varphi, \psi \in \Phi$ with $V\varphi \cap V\psi \neq \emptyset$. It now suffices to prove that $V\varphi \cup V\psi \subset U\varphi_0$ for some $\varphi_0 \in \Phi$. Choose $\varphi_0 \in \Phi$ so that $x_0\varphi_0 \in V\varphi \cap V\psi$. Since $x_0 \in V\varphi\varphi_0^{-1} \cap V\psi\varphi_0^{-1}$, $V\varphi\varphi_0^{-1} \cup V\psi\varphi_0^{-1} \subset U$ and $V\varphi \cup V\psi \subset U\varphi_0$. Hence $\mathfrak{U}$ is a uniformity of X. It is clear that Φ is uniformly equicontinuous relative to $\mathfrak{U}$. We show $\mathfrak{U}$ is compatible with the topology of X. If $x \in X$ and if $N \in \mathfrak{N}$, then $xN^* = \bigcup[N\varphi \mid \varphi \in \Phi, x \in N\varphi]$ is open. If $x \in X$ and if U is a neighborhood of x, then there exists $N \in \mathfrak{N}$ such that $\varphi \in \Phi$ with $x \in N\varphi$ implies $N\varphi \subset U$, whence $xN^* \subset U$. This completes the proof of the sufficiency.

Assume there exists a uniformity $\mathfrak{U}$ of X such that $\mathfrak{U}$ is compatible with the topology of X and Φ is uniformly equicontinuous relative to $\mathfrak{U}$. If $x \in X$ and if $\alpha \in \mathfrak{U}$, then there exist $\beta, \gamma \in \mathfrak{U}$ so that β is symmetric, $\beta^2 \subset \alpha$ and $x\gamma\varphi \subset$

$x\varphi\beta$ $(\varphi \in \Phi)$, whence $\varphi \in \Phi$ with $x \in x\gamma\varphi$ implies $x \in x\varphi\beta$, $x\varphi \in x\beta$, $x\gamma\varphi \subset x\varphi\beta \subset x\beta^2 \subset x\alpha$ and $x\gamma\varphi \subset x\alpha$. Let $x \in X$ and let $\mathfrak{N}$ be a neighborhood-base of x. There exists a base $\mathcal{V}$ of $\mathcal{U}$ such that $x\varphi\beta = x\beta\varphi$ $(\beta \in \mathcal{V})$. By 9.08, $[\bigcup_{y \in x\Phi} y\beta \times y\beta \mid \beta \in \mathcal{V}] = [\bigcup_{\varphi \in \Phi} x\beta\varphi \times x\beta\varphi \mid \beta \in \mathcal{V}]$ is a base of $\mathcal{U}$. Clearly, each of $[\bigcup_{\varphi \in \Phi} x\beta\varphi \times x\beta\varphi \mid \beta \in \mathcal{V}]$, $[\bigcup_{\varphi \in \Phi} N\varphi \times N\varphi \mid N \in \mathfrak{N}]$ refines the other. The proof is completed.

9.10. Remark. The following statements are pairwise equivalent:

(1) (X, T) is transitive; that is to say, if U and V are nonvacuous open subsets of X, then there exists $t \in T$ such that $Ut \cap V \neq \emptyset$.

(2) If U and V are nonvacuous open subsets of X, then $UT \cap V \neq \emptyset$.

(3) If U is a nonvacuous open subset of X, then $\overline{UT} = X$.

(4–5) Every invariant {nonvacuous open}{proper closed} subset of X is {everywhere}{nowhere} dense in X.

(6–7) If E and F are invariant {open}{closed} subsets of X such that $\{E \cap F = \emptyset\}\{E \cup F = X\}$, then $\{E = \emptyset$ or $F = \emptyset\}\{E = X$ or $F = X\}$.

(8–9) If E and F are invariant subsets of X such that $E \cap F = \emptyset$ and $E \cup F = X$, then {int $E = \emptyset$ or int $F = \emptyset$}$\{\overline{E} = X$ or $\overline{F} = X\}$.

(10) If E is an invariant subset of X, then int $E = \emptyset$ or $\overline{E} = X$.

9.11. Definition. A *Baire subset* of a topological space X is defined to be a subset E of X such that $E = U \Delta I$ for some open subset U of X and some first category subset I of X, where Δ denotes symmetric difference.

9.12. Theorem. *Let X be a complete metric space and let T be countable. Then the following statements are equivalent*:

(1) *(X, T) is transitive.*

(2) *If E is an invariant Baire subset of X, then either E or $X - E$ is a first category subset of X.*

Proof. Assume (1). We prove (2). Let E be an invariant second category Baire subset of X. It is enough to show that E is a residual subset of X. There exist an open subset U of X and a first category subset I of X such that $E = U \Delta I$ whence $U = E \Delta I = (E - I) \cup (I - E)$ and $UT = (E - \bigcap_{t \in T} It) \cup (I - E)T$. Thus $UT = (E - J) \cup K$ where J and K are first category subsets of X such that $J \subset E$ and $(E - J) \cap K = \emptyset$. It follows that $E = (UT - K) \cup J$. Since $U \neq \emptyset$, we have $\overline{UT} = X$. Hence UT is residual in X and E is residual in X. This proves that (1) implies (2).

Assume (2). We prove (1). Let U be an invariant nonvacuous open subset of X. It is enough to show that $\overline{U} = X$. Since U is a second category Baire subset of X, we have that U is residual in X and $\overline{U} = X$. The proof is completed.

9.13. Standing hypothesis. In 9.14–9.22 we assume that the phase group T is generative.

9.14. Remark. Let $x \in X$. Then the following statements are pairwise equivalent:

(1) (X, T) is extensively transitive at x; that is to say, if U is a nonvacuous open subset of X, then there exists an extensive subset A of T such that $xA \subset U$.

(2–3) If U is a nonvacuous open subset of X and if P is a replete semigroup in T, then $\{xP \cap U \neq \emptyset\}\{x \in UP\}$.

(4) If P is a replete semigroup in T, then $\overline{xP} = X$.

9.15. Remark. Let $x \in X$. Then (X, T) is extensively transitive at x if and only if (X, T) is both transitive and recurrent at x.

9.16. Remark. The transformation group (X, T) is pointwise extensively transitive if and only if (X, T) is both pointwise transitive and pointwise recurrent.

9.17. Remark. The following statements are pairwise equivalent:

(1) (X, T) is extensively transitive; that is to say, if U and V are nonvacuous open subsets of X, then there exists an extensive subset A of T such that $a \in A$ implies $Ua \cap V \neq \emptyset$.

(2) If U and V are nonvacuous open subsets of X and if P is a replete semigroup in T, then $UP \cap V \neq \emptyset$.

(3) If U is a nonvacuous open subset of X and if P is a replete semigroup in T, then $\overline{UP} = X$.

(4–5) If E is a {nonvacuous open}{proper closed} subset of X, if P is a replete semigroup in T and if $EP \subset E$, then E is {everywhere}{nowhere} dense in X.

(6–7) If E and F are {open}{closed} subsets of X such that $\{E \cap F = \emptyset\}$ $\{E \cup F = X\}$, if P is a replete semigroup in T, if $EP \subset E$ and if $FP^{-1} \subset F$, then $\{E = \emptyset$ or $F = \emptyset\}\{E = X$ or $F = X\}$.

(8–9) If E and F are subsets of X such that $E \cap F = \emptyset$ and $E \cup F = X$, if P is a replete semigroup in T and if $EP \subset E$ whence $FP^{-1} \subset F$, then {int $E = \emptyset$ or int $F = \emptyset\}\{\overline{E} = X$ or $\overline{F} = X\}$.

(10) If E is a subset of X, if P is a replete semigroup in T and if $EP \subset E$, then int $E = \emptyset$ or $\overline{E} = X$.

9.18. Remark. Certain universally valid implications among the transitivity properties are summarized in Table 4.

Table 4

T pointwise extensively transitive	→	T extensively transitive at x	→	T point extensively transitive	→	T extensively transitive
↓		↓		↓		↓
T pointwise transitive	→	T transitive at x	→	T point transitive	→	T transitive

9.19. Remark. Let Y be the set of all transitive points. Then:

(1) If $Y \neq \emptyset$, then (X, T) is transitive.

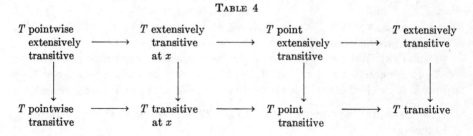

(2) Y is invariant.

(3) If $\mathcal{V}$ is a base of the topology of X such that $\emptyset \notin \mathcal{V}$, then $Y = \bigcap_{V \in \mathcal{V}} VT$.

(4) If X is second-countable, then Y is a G_δ subset of X.

(5) If X is second-countable and if (X, T) is transitive, then Y is a residual subset of X.

9.20. Theorem. *Let X be a complete separable metric space. Then the following statements are pairwise equivalent:*

(1) *(X, T) is transitive.*

(2) *(X, T) is point transitive.*

(3) *The set of all transitive points is an invariant residual G_δ subset of X.*

Proof. Use 9.19.

9.21. Remark. Let Z be the set of all extensively transitive points. Then:

(1) If $Z \neq \emptyset$, then T is extensively transitive.

(2) Z is invariant.

(3) If $\mathcal{V}$ is a base of the topology of X such that $\emptyset \notin \mathcal{V}$ and if $\mathcal{Q}$ is a base of the class of all replete semigroups in T, then $Z = \bigcap_{V \in \mathcal{V}} \bigcap_{Q \in \mathcal{Q}} VQ$.

(4) If X is second-countable, then Z is a G_δ subset of X.

(5) If X is second-countable and if T is extensively transitive, then Z is a residual subset of X.

9.22. Theorem. *Let X be a complete separable metric space. Then the following statements are pairwise equivalent:*

(1) *(X, T) is extensively transitive.*

(2) *(X, T) is point extensively transitive.*

(3) *The set of all extensively transitive points is an invariant residual G_δ subset of X.*

(4) *(X, T) is transitive and regionally recurrent.*

Proof. Use 9.21.

9.23. Theorem. *Let X be a complete separable metric space, let $T = \mathcal{J}$ or $\mathcal{R}$ and let every orbit under (X, T) be a first category subset of X. Then (X, T) is point extensively transitive if and only if (X, T) is transitive.*

Proof. Let (X, T) be transitive. Then (X, T) is regionally recurrent. Now apply 9.22.

9.24. Theorem. *Let X be second-countable, let n be a positive integer, let $(X, \mathcal{R}^n, \pi)$ be an extensively transitive n-parameter continuous flow and let E be the set of all $t \in \mathcal{R}^n$ such that the one-parameter discrete flow on X generated by π^t is extensively transitive. Then E is a residual G_δ subset of $\mathcal{R}^n$.*

Proof. Let $\mathcal{V}$ be a countable base of the topology of X such that $\emptyset \notin \mathcal{V}$. Let $\mathcal{P}$ be the class of all replete semigroups in $\mathcal{J}$. For $V, W \in \mathcal{V}$ and $P \in \mathcal{P}$ define $D(V, W, P)$ to be the set of all $t \in \mathcal{R}^n$ such that $V(p \cdot t) \cap W \neq \emptyset$ for some $p \in P$. Then $E = \bigcap_{V, W \in \mathcal{V}} \bigcap_{P \in \mathcal{P}} D(V, W, P)$. It remains to show that E

is a residual G_δ subset of $\mathcal{R}^n$. Let $V, W \in \mathcal{U}$ and let $P \in \mathcal{P}$. Clearly $D = D(V, W, P)$ is open in $\mathcal{R}^n$. Since $\mathcal{P}$ is countable, E is a G_δ subset of $\mathcal{R}^n$. We show D is dense in $\mathcal{R}^n$. Let U be a nonvacuous open subset of $\mathcal{R}^n$. Define $H = [p \cdot t \mid p \in P, t \in U]$. Since H contains a replete semigroup in $\mathcal{R}^n$, we have $VH \cap W \neq \emptyset$ whence $D \cap U \neq \emptyset$. The proof is completed.

9.25. Standing hypothesis. For the remainder of this section the condition that T is generative is omitted.

9.26. Definition. A homeomorphism φ of X onto X is said to be *motion preserving relative to* (X, T, π) or (X, T, π)-*motion preserving* provided that $x \in X$ implies $\pi_x \varphi = \pi_{x\varphi}$.

9.27. Remark. Let φ be a homeomorphism of X onto X. Then the following statements are pairwise equivalent:

(1) φ is motion preserving; that is to say, $\pi_x \varphi = \pi_{x\varphi}$ $(x \in X)$.
(2) $t\pi_x \varphi = t\pi_{x\varphi}$ $(x \in X, t \in T)$.
(3) $(x, t)\pi\varphi = (x\varphi, t)\pi$ $(x \in X, t \in T)$.
(4) $xt\varphi = x\varphi t$ $(x \in X, t \in T)$.
(5) $x\pi^t\varphi = x\varphi\pi^t$ $(x \in X, t \in T)$.
(6) $\pi^t\varphi = \varphi\pi^t$ $(t \in T)$.

9.28. Definition. The *motion preserving group of* (X, T, π) or the (X, T, π)-*motion preserving group* is defined to be the group of all motion preserving homeomorphisms relative to (X, T, π).

9.29. Definition. A homeomorphism φ of X onto X is said to be *orbit preserving relative to* (X, T, π) or (X, T, π)-*orbit preserving* provided that $x \in X$ implies $T\pi_x\varphi = T\pi_{x\varphi}$, or in different notation, $xT\varphi = x\varphi T$.

9.30. Definition. The *orbit preserving group* of (X, T, π) or the (X, T, π)-*orbit preserving group* is defined to be the group of all orbit preserving homeomorphisms relative to (X, T, π).

9.31. Definition. Let A be a subset of a group G. The {*centralizer*} {*normalizer*} *of* A *in* G is defined to be the group of all $x \in G$ such that {$a \in A$ implies $xa = ax$} {$xA = Ax$}.

9.32. Remark. Let Φ be the total homeomorphism group of X, let G be the transition group $[\pi^t \mid t \in T]$ of (X, T, π) and let {H} {K} be the {motion} {orbit} preserving group of (X, T, π). Then:

(1) H coincides with the centralizer of G in Φ.
(2) K contains the normalizer of G in Φ.
(3) $G \cup H \subset K$.
(4) If T is abelian, then $G \subset H$.
(5) If $G \subset H$ and if (X, T, π) is effective, then T is abelian.

9.33. Remark. Let X be a compact uniform space, let Φ be the total homeomorphism group of X, let Φ be provided with its space-index topology, let G be a subgroup of Φ and let H be the centralizer of G in Φ. Then:

(1) H is closed in Φ.

(2) If G is universally transitive abelian, then $G = H$.

(3) If G is equicontinuous transitive abelian, then $\overline{G} = H$ and H is equicontinuous universally transitive abelian.

9.34. Definition. Let $(\varphi_n \mid n \in I)$ be a sequence of functions on a topological space X to a uniform space Y, let φ be a function on X to Y and let $x \in X$. The sequence $(\varphi_n \mid n \in I)$ is said to *converge to* φ *uniformly at* x provided that if α is an index of Y, then there exist a neighborhood U of x and a positive integer N such that $y \in U$ and $n \in \mathscr{I}$ with $n \geqq N$ implies $(y\varphi, y\varphi_n) \in \alpha$.

9.35. Lemma. *Let X be a topological space, let Y be a metric space, let $(\varphi_n \mid n = 1, 2, \cdots)$ be a sequence of continuous functions on X to Y which converges pointwise to a function φ on X to Y and let E be the set of all $x \in X$ such that $(\varphi_n \mid n = 1, 2, \cdots)$ converges to φ uniformly at x. Then E is a residual G_δ subset of X.*

Proof. Let ρ be the metric of Y. For each positive number ϵ define $A(\epsilon)$ to be the set of all $x \in X$ such that $\rho(y\varphi, y\varphi_m) < \epsilon$ for all elements y of some neighborhood of x and all integers m greater than some positive integer. Clearly $A(\epsilon)$ is open in X for all positive numbers ϵ and $E = \bigcap_{n=1}^{+\infty} A(1/n)$.

Let ϵ be a positive number. It remains to show that $B = X - A(\epsilon)$ is a first category subset of X. Define $\delta = \epsilon/5$. For each positive integer n define C_n to be the set of all $x \in X$ such that $\rho(x\varphi, x\varphi_m) < \delta$ for all integers m such that $m \geqq n$. By hypothesis $X = \bigcup_{n=1}^{+\infty} C_n$ whence $B = \bigcup_{n=1}^{+\infty} B \cap C_n$. Let n be a positive integer and define $D = B \cap C_n$. The proof will be completed when we show that D is nowhere dense in X. Assume that D is somewhere dense in X, that is, there exists a nonvacuous open subset U of X such that $U \subset \overline{D}$. If $x \in U$ and if p is an integer such that $p \geqq n$, then there exists an integer q such that $q \geqq n$ and $\rho(x\varphi, x\varphi_q) < \delta$ and there exists $y \in U \cap D$ such that $\rho(x\varphi_p, y\varphi_p) < \delta$ and $\rho(x\varphi_q, y\varphi_q) < \delta$, whence $\rho(x\varphi, x\varphi_p) \leqq \rho(x\varphi, x\varphi_q) + \rho(x\varphi_q, y\varphi_q) + \rho(y\varphi_q, y\varphi) + \rho(y\varphi, y\varphi_p) + \rho(y\varphi_p, x\varphi_p) < 5\delta = \epsilon$. To summarize, if $x \in U$ and if p is an integer such that $p \geqq n$, then $\rho(x\varphi, x\varphi_p) < \epsilon$. By definition therefore, $U \subset A(\epsilon) = X - B$ whence $U \cap B = \emptyset$. However, since $U \subset \overline{D} \subset \overline{B}$ and U is nonvacuous open, $U \cap B \neq \emptyset$. This is a contradiction. The proof is completed.

9.36. Theorem. *Let X be a compact metric space, let Φ be the total homeomorphism group of X, let G be a universally transitive subgroup of Φ and let H be the centralizer of G in Φ. Then H is equicontinuous.*

Proof. Let $h_1, h_2, \cdots$ be a sequence in H. By 11.13 it is enough to show that $h_1, h_2, \cdots$ has a uniformly convergent subsequence. Choose $x_0 \in X$. Some subsequence of $x_0h_1, x_0h_2, \cdots$ converges. We may suppose $\lim_{n\to+\infty} x_0h_n$ exists; call this limit y_0. If $x \in X$ and if $g \in G$ such that $x_0g = x$, then $y_0g =$

$\lim_{n\to\infty} x_0 h_n g = \lim_{n\to\infty} x_0 g h_n = \lim_{n\to\infty} x h_n$. Hence $h_1, h_2, \cdots$ converges pointwise to some function h on X to X. By 9.35 there exists $x_1 \in X$ such that $h_1, h_2, \cdots$ converges to h uniformly at x_1.

We show $h_1, h_2, \cdots$ converges to h uniformly at all points of X. Let $x \in X$ and let ϵ be a positive number. There exists $g \in G$ such that $x_1 g = x$. Choose a positive number δ such that $z_1, z_2 \in X$ with $\rho(z_1, z_2) < \delta$ implies $\rho(z_1 g, z_2 g) < \epsilon$. There exists a neighborhood U_1 of x_1 and a positive integer N such that $\rho(zh, zh_n) < \delta$ for all $z \in U_1$ and all integers n with $n > N$, where ρ is the metric of X. Define $U = U_1 g$. Now U is a neighborhood of x. If $y \in U$ and if n is an integer with $n > N$, then $yg^{-1} \in U_1$ and $\rho(yhg^{-1}, yh_n g^{-1}) = \rho(yg^{-1}h, yg^{-1}h_n) < \delta$ whence $\rho(yh, yh_n) < \epsilon$. This shows that $h_1, h_2, \cdots$ converges to h uniformly at x.

It now follows that $h_1, h_2, \cdots$ converges uniformly to h. The proof is completed.

9.37. THEOREM. *Let X be a compact metric space, let T be abelian and let (X, T, π) be transitive. Then the following statements are equivalent:*

(1) *X is an almost periodic minimal orbit-closure under (X, T, π).*

(2) *The motion preserving group of (X, T, π) is universally transitive.*

PROOF. Use 4.35, 4.38, 9.33 and 9.36.

9.38. THEOREM. *Let X be a T_2-space, let X be minimal under T, let φ be a (X, T)-motion preserving homeomorphism such that $x\varphi \in xT$ for some $x \in X$ and let T be abelian. Then $\varphi = \pi^t$ for some $t \in T$.*

PROOF. Let $x \in X$ and $t \in T$ such that $x\varphi = x\pi^t$. Then $xs\varphi = xs\pi^t$ $(s \in T)$, $y\varphi = y\pi^t$ $(y \in xT)$ and $y\varphi = y\pi^t$ $(y \in X)$. The proof is completed.

9.39. NOTES AND REFERENCES.

(9.02) The term *transitive* was used by G. D. Birkhoff in 1920 (cf. Birkhoff [1], vol. 2, p. 108 and p. 221) to denote *regionally transitive* as defined here. The term is commonly used with the significance of the expression *universally transitive* as defined here.

(9.11) Cf., e.g., Kuratowski [1].

(9.12(2)) This is, in a sense, the topological analogue of metric transitivity. Cf. E. Hopf [2].

(9.17) Cf. Hilmy [2].

(9.24) Cf. Oxtoby and Ulam [2].

(9.35) Cf. Hausdorff [1], pp. 385–388.

(9.36) Cf. Gottschalk [9] and Fort [1].

Interesting examples of transitivity are to be found in Birkhoff [1], vol. 3, p. 307, Seidel and Walsh [1] and Oxtoby [1].

10. ASYMPTOTICITY

10.01. Standing hypothesis. In 10.02–10.10 we assume that X is a separated uniform space, and that (X, T, π) is a transformation group.

10.02. Definition. Let $x, y \in X$. The points x and y are said to be *separated* (each *from* the other) *under* T provided there exists an index α of X such that $t \in T$ implies $(xt, yt) \notin \alpha$. The points x and y are said to be *nonseparated* (each *from* the other) *under* T provided that $x \neq y$ and the pair x, y is not separated under T, that is, if α is an index of X, then there exists $t \in T$ such that $(xt, yt) \in \alpha$.

Let A and B be orbits under T. The orbits A and B are said to be {*separated*} {*nonseparated*} (each *from* the other) provided that $A \neq B$ and there exist $x \in A$ and $y \in B$ such that x and y are {separated} {nonseparated} under T.

10.03. Inheritance theorem. *Let X be compact, let $x, y \in X$ and let S be a syndetic subgroup of T. Then x and y are nonseparated under S if and only if x and y are nonseparated under T.*

Proof. Suppose x and y are nonseparated under T. Let α be an index of X and let K be a compact subset of T such that $T = SK$. There exists an index β of X such that $(x_1, x_2) \in \beta$ and $k \in K$ implies $(x_1k^{-1}, x_2k^{-1}) \in \alpha$. Select $t \in T$ such that $(xt, yt) \in \beta$. Now select $s \in S$ and $k \in K$ such that $t = sk$. It follows that $(xs, ys) = (xtk^{-1}, ytk^{-1}) \in \alpha$. The proof is completed.

10.04. Theorem. *Let X be a compact minimal orbit-closure under T and let $x, y \in X$ such that x and y are nonseparated under T. Then the T-traces of x and y coincide.*

Proof. Use 10.03 and 2.43.

10.05. Theorem. *Let X be a compact minimal orbit-closure under T and let T be regularly almost periodic at some point of X. Let $x, y \in X$ with $x \neq y$. Then the following statements are equivalent:*

(1) *x and y are nonseparated under T.*

(2) *The T-traces of x and y coincide.*

Proof. By 10.04, (1) implies (2). Assume (2). We prove (1). Let α be an index of X. Choose a symmetric index β of X such that $\beta^2 \subset \alpha$. Let T be regularly almost periodic at $z \in X$. There exists a closed syndetic invariant subgroup A of T such that $\overline{zA} \subset z\beta$. Let K be a compact subset of T such that $T = AK$. Since $X = \overline{zAK} = \bigcup_{k \in K} \overline{zkA}$, by 2.42 there exists $k \in K$ such that $x, y \in \overline{zkA}$ whence $xk^{-1}, yk^{-1} \in \overline{zA} \subset z\beta$ and $(xk^{-1}, yk^{-1}) \in \alpha$. The proof is completed.

10.06. Theorem. *Let X be compact, let X be minimal under T and let $x, y \in X$ such that $x \neq y$ and $x\varphi = y$ for some (X, T)-motion preserving homeomorphism φ. Then x and y are separated under T.*

PROOF. Assume x and y are nonseparated. Choose open neighborhoods U, V of x, y such that $U \cap V = \emptyset$ and $U\varphi \subset V$. There exists a finite subset E of T such that $X = UE$. Select $t \in T$ and $s \in E$ such that xt, $yt \in Us$ whence xts^{-1}, $yts^{-1} \in U$. It follows that $yts^{-1} = x\varphi ts^{-1} = xts^{-1}\varphi \in U\varphi \subset V$ and $yts^{-1} \in U \cap V$. This is a contradiction. The proof is completed.

10.07. THEOREM. *Let X be compact, let X be minimal under T, let x, $y \in X$ such that $x \neq y$ and $y \in xT$, and let T be abelian. Then x and y are separated under T.*

PROOF. Use 9.32(4) and 10.06.

10.08. THEOREM. *Let X be locally compact, let $x, y \in X$, let both x and y be almost periodic under T, let x and y be nonseparated under T and let T be locally compact. Then neither x nor y is regularly almost periodic under T.*

PROOF. Assume x is regularly almost periodic under T. Then there exists a closed syndetic invariant subgroup S of T such that $y \notin \overline{xS}$. Since x and y are almost periodic under S, we have that $\overline{xS} \cap \overline{yS} = \emptyset$. It follows from 10.03 that the pair x, y is separated under T. This contradicts the hypothesis and the proof is completed.

10.09. REMARK. Let X be compact, let x, $y \in X$ and let φ be a (X, T)-motion preserving homeomorphism. Then $x\varphi$ and $y\varphi$ are separated under T if and only if x and y are separated under T.

10.10. THEOREM. *Let X be compact, let X be minimal under T, let there exist exactly one nonseparated pair of orbits under T and let T be abelian. Then every (X, T)-motion preserving homeomorphism is a transition of (X, T).*

PROOF. Let x, $y \in X$ such that x, y are nonseparated and $xT \neq yT$. Let φ be a motion preserving homeomorphism. Since $x\varphi T$, $y\varphi T$ are nonseparated orbits, we have $x\varphi T = xT$ or $x\varphi T = yT$. It is enough by 9.38 to show that $x\varphi T = xT$. Assume $x\varphi T = yT$. Then $x\varphi\pi^t = y$ for some $t \in T$. By 10.06, x and y are separated. This is a contradiction. The proof is completed.

10.11. STANDING NOTATION. Throughout the remainder of this section X denotes a compact metric space with metric ρ and φ denotes a homeomorphism of X onto X. It is somewhat more convenient to speak of the homeomorphism φ than the discrete flow generated by φ.

10.12. DEFINITION. Let $x \in X$. The *orbit of* x, denoted $O(x)$, is the set $\bigcup_{n=-\infty}^{+\infty} x\varphi^n$. The {*negative*}{*positive*} *semiorbit of* x, denoted $\{O^-(x)\}\{O^+(x)\}$, is the set $\{\bigcup_{n=0}^{-\infty} x\varphi^n\}\{\bigcup_{n=0}^{+\infty} x\varphi^n\}$. To indicate dependence on φ we may suffix the phrase "under φ" and adjoin a subscript φ to O.

10.13. REMARK. Let $x \in X$. Then the negative semiorbit of x under φ coincides with the positive semiorbit of x under φ^{-1}.

10.14. Remark. Let $x \in X$. Then:

(1) $O(x) = O^-(x) \cup O^+(x)$.
$O(x)\varphi^{-1} = O(x\varphi^{-1}) = O(x) = O(x\varphi) = O(x)\varphi$.
$O^-(x)\varphi^{-1} = O^-(x\varphi^{-1}) \subset O^-(x) \subset O^-(x\varphi) = O^-(x)\varphi$.
$O^+(x)\varphi^{-1} = O^+(x\varphi^{-1}) \supset O^+(x) \supset O^+(x\varphi) = O^+(x)\varphi$.

(2) $\overline{O}(x) = \overline{O}^-(x) \cup \overline{O}^+(x)$.
$\overline{O}(x)\varphi^{-1} = \overline{O}(x\varphi^{-1}) = \overline{O}(x) = \overline{O}(x\varphi) = \overline{O}(x)\varphi$.
$\overline{O}^-(x)\varphi^{-1} = \overline{O}^-(x\varphi^{-1}) \subset \overline{O}^-(x) \subset \overline{O}^-(x\varphi) = \overline{O}^-(x)\varphi$.
$\overline{O}^+(x)\varphi^{-1} = \overline{O}^+(x\varphi^{-1}) \supset \overline{O}^+(x) \supset \overline{O}^+(x\varphi) = \overline{O}^+(x)\varphi$.

10.15. Definition. Let $x \in X$. The *{α-limit}{ω-limit} set of x*, denoted $\{\alpha(x)\}\{\omega(x)\}$, is the set of all $y \in X$ such that $\lim_{i\to+\infty} x\varphi^{n_i} = y$ for some sequence $n_1, n_2, \cdots$ of integers such that $\{n_1 > n_2 > \cdots\}\{n_1 < n_2 < \cdots\}$. Each point of $\{\alpha(x)\}\{\omega(x)\}$ is called an *{α-limit}{ω-limit} point of x*. To indicate dependence on φ we may suffix the phrase "under φ" and adjoin a subscript φ to α and ω. This definition agrees with 6.33.

10.16. Remark. Let $x \in X$. Then the α-limit set of x under φ coincides with the ω-limit set of x under φ^{-1}.

10.17. Remark. Let $x \in X$. Then:

(1) $\overline{O}(x) = O(x) \cup \alpha(x) \cup \omega(x)$.
$\overline{O}^-(x) = O^-(x) \cup \alpha(x)$.
$\overline{O}^+(x) = O^+(x) \cup \omega(x)$.

(2) $\alpha(x)$ and $\omega(x)$ are nonvacuous closed invariant.

(3) $\alpha(x\varphi^{-1}) = \alpha(x) = \alpha(x\varphi)$.
$\omega(x\varphi^{-1}) = \omega(x) = \omega(x\varphi)$.

(4) $\alpha(x) = \bigcap_{n=0}^{-\infty} (\bigcup_{m=n}^{-\infty} x\varphi^m)^-$.
$\omega(x) = \bigcap_{n=0}^{\infty} (\bigcup_{m=n}^{+\infty} x\varphi^m)^-$.

(5) Every point of $\alpha(x) \cup \omega(x)$ is regionally recurrent.

10.18. Definition. Let $x \in X$. The homeomorphism φ is said to be *{negatively}{positively} recurrent at x* and the point x is said to be *{negatively}{positively} recurrent under* φ provided there exists a sequence $n_1, n_2, \cdots$ of integers such that $\{n_1 > n_2 > \cdots\}\{n_1 < n_2 < \cdots\}$ and $\lim_{i\to+\infty} x\varphi^{n_i} = x$. The homeomorphism φ is said to be *recurrent at x* and the point x is said to be *recurrent under* φ provided that φ is both negatively and positively recurrent at x. This definition agrees with 3.36.

10.19. Remark. Let $x \in X$. Then x is negatively recurrent under φ if and only if x is positively recurrent under φ^{-1}.

10.20. Remark. Let $x \in X$. Then:

(1) The following statements are pairwise equivalent:

(I) x is negatively recurrent.

(II) $x \in \alpha(x)$.

(III) $x \in \bar{O}^-(x\varphi^{-1})$.
(IV) $\bar{O}(x) = \bar{O}^-(x) = \alpha(x) \supset \omega(x)$.
(2) The following statements are pairwise equivalent:
(I) x is positively recurrent.
(II) $x \in \omega(x)$.
(III) $x \in \bar{O}^+(x\varphi)$.
(IV) $\bar{O}(x) = \bar{O}^+(x) = \omega(x) \supset \alpha(x)$.
(3) The following statements are pairwise equivalent:
(I) x is recurrent.
(II) $x \in \alpha(x) \cap \omega(x)$.
(III) $x \in \bar{O}^-(x\varphi^{-1}) \cap \bar{O}^+(x\varphi)$.
(IV) $\bar{O}(x) = \bar{O}^-(x) = \bar{O}^+(x) = \alpha(x) = \omega(x)$.

10.21. Definition. Let $x \in X$ and let B be a closed invariant subset of X. The point x is said to be {*negatively*}{*positively*} *asymptotic to* B *under* φ provided that $x \notin B$ and $\{\lim_{n\to-\infty} \rho(x\varphi^n, B) = 0\}\{\lim_{n\to+\infty} \rho(x\varphi^n, B) = 0\}$.

Let A be an orbit under φ and let B be a closed invariant subset of X. The orbit A is said to be {*negatively*}{*positively*} *asymptotic to* B provided there exists $x \in A$ such that x is {negatively}{positively} asymptotic to B under φ.

Let $x, y \in X$. The points x and y are said to be {*negatively*}{*positively*} *asymptotic* (each *to* the other) *under* φ provided that $x \neq y$ and $\{\lim_{n\to-\infty} \rho(x\varphi^n, y\varphi^n) = 0\}$ $\{\lim_{n\to+\infty} (x\varphi^n, y\varphi^n) = 0\}$.

Let A and B be orbits under φ. The orbits A and B are said to be {*negatively*} {*positively*} *asymptotic* (each *to* the other) provided that $A \neq B$ and there exist $x \in A$ and $y \in B$ such that x and y are {negatively}{positively} asymptotic under φ.

The term {*asymptotic*}{*doubly asymptotic*} means negatively {or}{and} positively asymptotic.

10.22. Remark. Let $x \in X$ and let A be an invariant closed subset of X. Then x is negatively asymptotic to A under φ if and only if x is positively asymptotic to A under φ^{-1}.

Let $x, y \in X$. Then x and y are negatively asymptotic under φ if and only if x and y are positively asymptotic under φ^{-1}.

10.23. Remark. Let $x \in X$, let A be an invariant closed subset of X and let n be an integer. Then $x\varphi^n$ is {negatively}{positively} asymptotic to A under φ if and only if x is {negatively}{positively} asymptotic to A under φ.

Let $x, y \in X$ and let n be an integer. Then $x\varphi^n$ and $y\varphi^n$ are {negatively} {positively} asymptotic under φ if and only if x and y are {negatively}{positively} asymptotic under φ.

10.24. Remark. Let $x \in X$. Then:
(1) x is negatively asymptotic to $\alpha(x)$ if and only if $x \notin \alpha(x)$.
(2) x is positively asymptotic to $\omega(x)$ if and only if $x \notin \omega(x)$.
(3) x is doubly asymptotic to $\alpha(x) \cup \omega(x)$ if and only if $x \notin \alpha(x) \cup \omega(x)$.

10.25. Remark. Let A be the set of all regionally recurrent points of X. Then A is a nonvacuous closed invariant subset of X such that every point of $X - A$ is doubly asymptotic to A.

10.26. Inheritance theorem. *Let n be a positive integer. Then:*

(1) *If $x \in X$ and if A is an invariant closed subset of X, then x is {negatively} {positively} asymptotic to A under φ^n if and only if x is {negatively} {positively} asymptotic to A under φ.*

(2) *If $x, y \in X$, then x and y are {negatively} {positively} asymptotic under φ^n if and only if x and y are {negatively} {positively} asymptotic under φ.*

Proof. Obvious.

10.27. Theorem. *Let $x \in X$ and let A be a closed invariant subset of X such that $x \notin A$. Then the following statements are pairwise equivalent:*

(1) *x is {negatively} {positively} asymptotic to A.*

(2) *$\{\alpha(x) \subset A\}\{\omega(x) \subset A\}$.*

(3) *If U is a neighborhood of A, then there exists an integer n such that $\{\bigcup_{m=n}^{-\infty} x\varphi^m \subset U\}\{\bigcup_{m=n}^{+\infty} x\varphi^m \subset U\}$.*

Proof. Obvious.

10.28. Theorem. *Let A be a closed non-open invariant subset of X and let there exist a neighborhood U of A such that $x \in U - A$ implies $O(x) \not\subset U$. Then there exists $y \in X$ such that y is asymptotic to A.*

Proof. Let V be an open neighborhood of A such that $\overline{V} \subset U$. Choose a sequence $x_1, x_2, \cdots$ of points of $X - A$ such that $\lim_{i \to +\infty} x_i \in A$. For each positive integer i let n_i be the integer with least absolute value such that $x_i\varphi^{n_i} \in X - V$. We may assume that $0 \le n_1 \le n_2 \cdots$ and $\lim_{i \to +\infty} x_i\varphi^{n_i} = y \in X - V$. It follows that $\bigcup_{n=-1}^{-\infty} y\varphi^n \subset \overline{V}$. By 10.24(1) it is enough to show that $\alpha(y) \subset A$. Assume there exists $x \in \alpha(y) - A$. Then $O(x) \subset \alpha(y) \subset \overline{V} \subset U$. This contradicts the hypothesis. The proof is completed.

10.29. Theorem. *Let A be a subset of X such that $A\varphi \subset A$ and let U be an open neighborhood of A. Then at least one of the following statements is valid:*

(1) *There exists a closed subset E of X such that $A \subset E \not\subset U$, $E\varphi \subset E$ and $E \subset \overline{U}\varphi^{-1}$.*

(2) *There exists an open subset V of X such that $A \subset V \subset U$ and $V\varphi^{-1} \subset V$.*

Proof. Define $E = \bigcap_{n=1}^{+\infty} \overline{U}\varphi^{-n}$. Now E is closed, $E\varphi \subset E$ and $E \subset \overline{U}\varphi^{-1}$. Since $A \subset A\varphi^{-1} \subset A\varphi^{-2} \subset \cdots$, we have $A \subset A\varphi^{-n} \subset U\varphi^{-n}$ $(n = 1, 2, \cdots)$ and $A \subset E$. If $E \not\subset U$, then (1) holds. Assume $E \subset U$. There exists a positive integer m such that $\bigcap_{n=1}^{m} \overline{U}\varphi^{-n} \subset U$. Define $V = \bigcap_{n=1}^{m} U\varphi^{-n}$. Now V is open and $A \subset V \subset U$. Since $V = U \cap \bigcap_{n=1}^{m} U\varphi^{-n}$, we have $V\varphi^{-1} = \bigcap_{n=1}^{m+1} U\varphi^{-n} \subset V$. The proof is completed.

10.300. Lemma. *Let X be a uniform space, let Φ be an abelian set of permutations of X, let $h: \Phi \to \Phi^{-1}$ be the inversion map $\varphi \to \varphi^{-1}$, and let Φ, Φ^{-1} each be* *provided with its space-index uniformity. Then:*

(1) *h is a unimorphism of Φ onto Φ^{-1}.*

(2) *Φ is totally bounded if and only if Φ^{-1} is totally bounded.*

10.30. **Theorem.** *Let X be infinite and let ϵ be a positive number. Then there exist $x, y \in X$ with $x \neq y$ such that $n \geqq 1$ implies $\rho(x\varphi^n, y\varphi^n) < \epsilon$.*

Proof. Define the homeomorphism

$$\psi : X \times X \overset{\text{onto}}{\rightarrow} X \times X$$

by $(x, y)\psi = (x\varphi, y\varphi)$ $(x, y \in X)$. The diagonal A of $X \times X$ is a closed non-open ψ-invariant subset of $X \times X$. For n a positive integer, define $U_n = \{(x, y) \mid x, y \in X, \rho(x, y) < 1/n\}$. Clearly U_n is an open subset of $X \times X$ and $A \subset U_n$. Now apply 10.29. If conclusion (1) holds for infinitely many positive integers n, the proof of the theorem is completed. If this is not the case, there exists a positive integer N such that $n > N$ implies the existence of an open subset V_n of $X \times X$ such that $A \subset V_n \subset U_n$ and $V_n\psi^{-1} \subset V_n$. It follows that the set $\Phi^- = [\varphi^{-i} \mid i = 0, 1, 2, \cdots]$ is an equicontinuous set of homeomorphisms of X onto X, and thus, by 11.12, Φ is totally bounded in its compact index uniformity. Let $\Phi^+ = [\varphi^i \mid i = 0, 1, 2, \cdots]$. By 10.300, the set Φ^+ is totally bounded in its space index uniformity, hence equicontinuous, and corresponding to $\epsilon > 0$ there exists $\delta > 0$ such that $x, y \in X$ with $\rho(x, y) < \delta$ and $n \geqq 1$ imply $\rho(x\varphi^n, y\varphi^n) < \epsilon$. Since X is infinite we can determine $x, y \in X$ with $x \neq y$ and $\rho(x, y) < \delta$. The proof of the theorem is completed.

10.31. Definition. The homeomorphism φ is said to be *expansive* provided there exists a positive number d such that if $x, y \in X$ with $x \neq y$, then there exists an integer n such that $\rho(x\varphi^n, y\varphi^n) > d$.

Let $x \in X$. The homeomorphism φ is said to be *expansive at x*, and the point x is said to be *expansive under φ* provided there exists a positive number d such that if $y \in X$ with $y \neq x$, then there exists an integer n such that $\rho(x\varphi^n, y\varphi^n) > d$.

10.32. Inheritance theorem. *Let n be a nonzero integer. Then:*

(1) *φ^n is expansive if and only if φ is expansive.*

(2) *If $x \in X$, then φ^n is expansive at x if and only if φ is expansive at x.*

Proof. (1) Let φ be expansive. We show φ^n is expansive. We may assume $n > 0$. There exists a positive number d such that $x, y \in X$ with $x \neq y$ implies $\rho(x\varphi^m, y\varphi^m) > d$ for some integer m. Choose a positive number c such that $x, y \in X$ with $\rho(x, y) > d$ implies $\rho(x\varphi^i, y\varphi^i) > c$ $(i = 0, \cdots, n - 1)$. Let $x, y \in X$ with $x \neq y$. There exists an integer m such that $\rho(x\varphi^m, y\varphi^m) > d$. Select integers p, r such that $np = m + r$ and $0 \leqq r < n$. It follows that $\rho(x\varphi^{np}, y\varphi^{np}) = \rho(x\varphi^{m+r}, y\varphi^{m+r}) > c$. This proves that φ^n is expansive.

The converse is obvious. The proof is completed.

(2) Similar to (1).

10.33. DEFINITION. Let $x \in X$. The homeomorphism φ is said to be *periodic at* x and the point x is said to be *periodic under* φ provided there exists a positive integer n such that $x\varphi^n = x$. If φ is periodic at x, then the *period of* φ *at* x and the *period of* x under φ is defined to be the least positive integer n such that $x\varphi^n = x$. The homeomorphism φ is said to be *fixed at* x and the point x is said to be *fixed under* φ provided that $x\varphi = x$, that is, the period of φ at x is 1.

The homeomorphism φ is said to be *pointwise periodic* provided that φ is periodic at every point of X.

The homeomorphism φ is said to be *periodic* provided there exists a positive integer n such that $x \in X$ implies $x\varphi^n = x$. If φ is periodic, then the *period of* φ is defined to be the least positive integer n such that $x \in X$ implies $x\varphi^n = x$.

It is clear that this definition agrees essentially with 3.06.

10.34. THEOREM. *Let* φ *be expansive. Then*:

(1) *If* n *is a positive integer, then the set of all points of* X *with period* n *is finite.*

(2) *The set of all periodic points of* X *is countable.*

PROOF. It is clear that the set of all fixed points is finite. Now use 10.32.

10.35. THEOREM. *Let* X *be self-dense and let* φ *be expansive. Then*:

(1) φ *is not pointwise periodic.*

(2) φ *is not almost periodic.*

PROOF. To prove (1) use 10.34(1). To prove (2) use 4.35 and 4.37.

10.36. THEOREM. *Let* X *be infinite and let* φ *be expansive. Then there exists a pair of points of* X *which are positively asymptotic under* φ *and a pair of points of* X *which are negatively asymptotic under* φ.

PROOF. It is sufficient to prove the existence of a pair of points of X which are positively asymptotic under φ. Since φ is expansive, there exists a positive number d such that if $x, y \in X$ with $x \neq y$, then there exists an integer m such that $\rho(x\varphi^m, y\varphi^m) > d$. According to 10.30, there exist $x_0, y_0 \in X$ with $x_0 \neq y_0$ such that $n \geqq 1$ implies $\rho(x_0\varphi^n, y_0\varphi^n) < d/2$. We show that x_0 and y_0 are positively asymptotic under φ. If this is not the case, there exists a positive number ϵ and a sequence of integers $n_1 < n_2 < \cdots$ such that $i \geqq 1$ implies $\rho(x_0\varphi^{n_i}; y_0\varphi^n) > \epsilon$. We can assume that $\lim_{i\to\infty} x_0\varphi^{n_i} = x_1$ and $\lim_{i\to\infty} y_0\varphi^{n_i} = y_1$. Then $\rho(x_1, y_1) \geqq \epsilon$ and thus $x_1 \neq y_1$. Let k be any integer. Then

$$\lim_{i\to\infty} x_0\varphi^{n_i+k} = x_1\varphi^k, \ \lim_{i\to\infty} y_0\varphi^{n_i+k} = y_1\varphi^k,$$

and since for i sufficiently large, $n_i + k > 0$, it follows that $\rho(x_1\varphi^k, y_1\varphi^k) < d$, contrary to the hypothesis that ϕ is expansive.

The proof of the theorem is completed.

10.37. THEOREM. *Let* $x \in X$, *let* x *be non-isolated and let* φ *be periodic at* x *and expansive at* x. *Then there exists* $y \in X$ *such that* y *is asymptotic to* x.

PROOF. Use 10.32, 10.28 and 10.26.

10.38. Definition. Let $x, y \in X$. The points x and y are said to be {*negatively*}{*positively*} *separated* (each *from* the other) *under* φ provided there exists a positive number ϵ such that if n is a {negative}{positive} integer, then $\rho(x\varphi^n, y\varphi^n) \geqq \epsilon$. The points x and y are said to be {*negatively*}{*positively*} *nonseparated* (each *from* the other) *under* φ provided that $x \neq y$ and if ϵ is a positive number, then there exists a {negative}{positive} integer n such that $\rho(x\varphi^n, y\varphi^n) < \epsilon$.

The term *separated* means negatively and positively separated. The term *nonseparated* means negatively or positively nonseparated. The term *doubly nonseparated* means negatively and positively nonseparated.

Let A and B be orbits under φ. The orbits A and B are said to be *admissible* (each *from* the other) provided that $A \neq B$ and there exist $x \in A$ and $y \in B$ such that x and y are admissible under φ, where "admissible" is replaceable by one of the seven terms defined above.

This definition agrees with 10.02.

10.39. Remark. Let $x, y \in X$. Then:

(1) x and y are negatively nonseparated under φ if and only if x and y are positively nonseparated under φ^{-1}.

(2) If n is an integer, then $x\varphi^n$ and $y\varphi^n$ are {negatively}{positively} nonseparated under φ if and only if x and y are {negatively}{positively} nonseparated under φ.

(3) If x and y are {negatively}{positively} asymptotic under φ, then x and y are {negatively}{positively} nonseparated under φ.

10.40. Inheritance theorem. *Let $x, y \in X$ and let n be a positive integer. Then x and y are {negatively}{positively} nonseparated under φ^n if and only if x and y are {negatively}{positively} nonseparated under φ.*

Proof. Obvious.

10.41. Theorem. *Let X be a locally recurrent minimal orbit-closure and let $x, y \in X$. Then x and y are negatively nonseparated if and only if x and y are positively nonseparated.*

Proof. Let x and y be negatively nonseparated. It is enough to show that x and y are positively nonseparated. Let ϵ be a positive number. There exist a positive number δ and an extensive subset E of $\mathcal{I}$ such that $\bigcup_{n\in E} x\delta\varphi^n \subset xE$. Choose a finite subset F of $\mathcal{I}$ such that $X = \bigcup_{m\in F} x\delta\varphi^m$. There exists a negative integer k such that $x\varphi^k, y\varphi^k \in x\delta\varphi^m$ for some $m \in F$ whence $x\varphi^{k-m}, y\varphi^{k-m} \in x\delta$. Now select $n \in E$ such that $p = k - m + n > 0$. It follows that $x\varphi^p, y\varphi^p \in x\epsilon$ and $\rho(x\varphi^p, y\varphi^p) < 2\epsilon$. The proof is completed.

10.42. Theorem. *Let X be a minimal orbit-closure and let x, y be a nonseparated pair of points of X. Then neither x nor y is regularly almost periodic.*

Proof. Use 10.08.

10.43. THEOREM. *Let X be a minimal orbit-closure. Then:*

(1) *φ is weakly almost periodic.*

(2) *The following statements are pairwise equivalent:*

(I) *Some point of X is locally almost periodic.*

(II) *φ is locally almost periodic.*

(III) *φ is weakly$_2$ almost periodic.*

(3) *The following statements are pairwise equivalent:*

(I) *Some point of X is isochronous.*

(II) *φ is weakly isochronous.*

(III) *The set of all regularly almost periodic points of X is an invariant residual G_δ subset of X.*

(4) *If φ is weakly isochronous, then φ is locally almost periodic.*

(5) *φ is regularly almost periodic if and only if φ is pointwise regularly almost periodic.*

PROOF. Use 4.24, 4.31, 5.09, 5.23, 5.24, 5.18.

10.44. THEOREM. *Let X be a weakly isochronous minimal orbit-closure and let A be the set of all $x \in X$ such that x is nonseparated from some point of X. Then A is an invariant first category subset of X.*

PROOF. Use 10.42 and 10.43.

10.45. NOTES AND REFERENCES.

(10.15) This terminology is due to G. D. Birkhoff [2].

(10.29) This theorem is due to Montgomery [3]. See also Kerékjártó [1].

(10.30 and 10.36) These theorems are due to S. Schwartzman [1]. A weaker form of 10.36 was proved by Utz [1].

(10.32, 10.34, 10.37) Cf. Utz [1].

(10.02, 10.38) The terms *distal* and *proximal* for separated and nonseparated, respectively, have been proposed and are being used by Gottschalk.

11. FUNCTION SPACES

11.01. Definition. Let X be a set, let Y be a uniform space with uniformity I and let Φ be a set of mappings of X into Y. For $\alpha \subset Y \times Y$ define $\alpha_\Phi = [(\varphi, \psi) \mid \varphi, \psi \in \Phi \;\&\; (x\varphi, x\psi) \in \alpha \; (x \in X)]$. Define

$$\mathcal{V} = [\alpha_\Phi \mid \alpha \in I].$$

It is readily verified that $\mathcal{V}$ is a uniformity-base of Φ. The uniformity $\mathcal{U}$ of Φ generated by $\mathcal{V}$ is called the *space-index* uniformity of Φ. The topology of Φ induced by $\mathcal{U}$ is called the *space-index* topology of Φ.

11.02. Theorem. *Let X be a topological space, let Y be a uniform space, let Φ be a set of continuous mappings of X into Y, let Φ be provided with its space-index topology and let $\pi : X \times \Phi \to Y$ be defined by $(x, \varphi)\pi = x\varphi$ $(x \in X, \varphi \in \Phi)$. Then π is continuous.*

Proof. Let $x \in X$, let $\varphi \in \Phi$ and let V be a neighborhood of $(x, \varphi)\pi = x\varphi$. Choose an index β of Y and a neighborhood U of x such that $x\varphi\beta^2 \subset V$ and $U\varphi \subset x\varphi\beta$. We show $(U \times \varphi\beta_\Phi)\pi \subset V$. Let $y \in U$ and let $\psi \in \varphi\beta_\Phi$. Then $y\varphi \in x\varphi\beta$, $(x\varphi, y\varphi) \in \beta$, $(\varphi, \psi) \in \beta_\Phi$, $(y\varphi, y\psi) \in \beta$, $(x\varphi, y\psi) = (x\varphi, y\varphi) \cdot (y\varphi, y\psi) \in \beta^2$ and $(y, \psi)\pi = y\psi \in x\varphi\beta^2 \subset V$. The proof is completed.

11.03. Theorem. *Let X be a set, let Y be a complete uniform space and let Φ be the set of all mappings of X into Y. Then Φ is complete in its space-index uniformity.*

Proof. Let Φ be provided with its space-index uniformity. Let $\mathfrak{F}$ be a cauchy filter on Φ. For each $x \in X$, $[xF \mid F \in \mathfrak{F}]$ is a cauchy filter-base on Y. Define $\varphi : X \to Y$ by $x\varphi \in \bigcap_{F \in \mathfrak{F}} \overline{xF}$ $(x \in X)$. We show $\mathfrak{F} \to \varphi$. Let α be a closed index of Y. There exists $F \in \mathfrak{F}$ such that $F \times F \subset \alpha_\Phi$. Then $xF \times xF \subset \alpha$ $(x \in X)$, $\overline{xF} \times \overline{xF} \subset \alpha$ $(x \in X)$, $[x\varphi] \times xF \subset \alpha$ $(x \in X)$, $[\varphi] \times F \subset \alpha_\Phi$ and $F \subset \varphi\alpha_\Phi$. The proof is completed.

11.04. Theorem. *Let X be a {topological} {uniform} space, let Y be a uniform space, let Φ be the set of all mappings of X into Y, let Φ be provided with its space-index topology and let Ψ be the set of all {continuous} {uniformly continuous} mappings of X into Y. Then Ψ is a closed subset of Φ.*

Proof. First reading. Let $\varphi \in \overline{\Psi}$. We show φ is continuous. Let $x_0 \in X$ and let α be an index of Y. Choose a symmetric index β of Y such that $\beta^3 \subset \alpha$. There exists $\psi \in \Psi \cap \varphi\beta_\Phi$. Select a neighborhood U of x_0 such that $U\psi \subset x_0\psi\beta$. If $x \in U$, then $(x_0\varphi, x_0\psi) \in \beta$, $(x_0\psi, x\psi) \in \beta$, $(x\psi, x\varphi) \in \beta$, $(x_0\varphi, x\varphi) \in \beta^3 \subset \alpha$ and $x\varphi \subset x_0\varphi\alpha$. The proof is completed.

Second reading. Let $\varphi \in \overline{\Psi}$. We show φ is uniformly continuous. Let β be an index of Y. Choose a symmetric index γ of Y such that $\gamma^3 \subset \beta$. There exists

$\psi \in \Psi \cap \varphi\gamma_\Phi$. Select an index α of X such that $(x_1, x_2) \in \alpha$ implies $(x_1\psi, x_2\psi) \in \gamma$. If $(x_1, x_2) \in \alpha$, then $(x_1\varphi, x_1\psi) \in \gamma$, $(x_1\psi, x_2\psi) \in \gamma$, $(x_2\psi, x_2\varphi) \in \gamma$ and $(x_1\varphi, x_2\varphi) \in \gamma^3 \subset \beta$. The proof is completed.

11.05. THEOREM. *Let X be a {topological} {uniform} space, let Y be a complete uniform space and let Φ be the set of all {continuous} {uniformly continuous} mappings of X into Y. Then Φ is complete in its space-index uniformity.*

PROOF. Use 11.03 and 11.04.

11.06. THEOREM. *Let X be a set, let Y be a totally bounded uniform space and let Φ be a set of mappings of X into Y. Then the following statements are equivalent:*

(1) *Φ is totally bounded in its space-index uniformity.*

(2) *If α is an index of Y, then there exists a finite partition $\mathfrak{A}$ of X such that $A \in \mathfrak{A}$ and $\varphi \in \Phi$ implies $A\varphi \times A\varphi \subset \alpha$.*

PROOF. We show that (1) implies (2). Assume (1). Let α be an index of Y. Choose a symmetric index β of Y such that $\beta^3 \subset \alpha$. There exists a finite subset F of Φ such that $\Phi = \bigcup_{f \in F} f\beta_\Phi$. Since Y is totally bounded, for each $f \in F$ there exists a finite partition $\mathfrak{A}_f$ of X such that $Af \times Af \subset \beta$ $(A \in \mathfrak{A}_f)$. Define $\mathfrak{A} = \dot{\bigcap}_{f \in F} \mathfrak{A}_f$. Clearly $\mathfrak{A}$ is a finite partition of X such that $Af \times Af \subset \beta$ $(A \in \mathfrak{A}, f \in F)$. Let $A \in \mathfrak{A}$ and let $\varphi \in \Phi$. We show $A\varphi \times A\varphi \subset \alpha$. Choose $f \in F$ such that $\varphi \in f\beta_\Phi$. Then $x\varphi \in xf\beta$ $(x \in A)$, $A\varphi \subset Af\beta$, $A\varphi \times A\varphi \subset Af\beta \times Af\beta = \beta(Af \times Af)\beta \subset \beta^3 \subset \alpha$. Hence (1) implies (2).

We show that (2) implies (1). Assume (2). Let α be an index of Y. Choose a symmetric index β of Y such that $\beta^4 \subset \alpha$. There exists a finite partition $\mathfrak{A}$ of X such that $A\varphi \times A\varphi \subset \beta$ $(A \in \mathfrak{A}, \varphi \in \Phi)$. Select a finite subset E of Y such that $Y = E\beta$. If $A \in \mathfrak{A}$ and if $\varphi \in \Phi$, then there exists $y \in E$ such that $y\beta \cap A\varphi \neq \emptyset$ whence $A\varphi = y\beta(A\varphi \times A\varphi) \subset y\beta^2$. That is to say, if $\varphi \in \Phi$, then to each $A \in \mathfrak{A}$ there corresponds at least one $y \in E$ such that $A\varphi \subset y\beta^2$. Each $\varphi \in \Phi$ thus determines a nonvacuous set φ^* of mappings of $\mathfrak{A}$ into E as follows: $t \in \varphi^*$ if and only if $A\varphi \subset At\beta^2$ $(A \in \mathfrak{A})$. Since the set of all mappings of $\mathfrak{A}$ into E is finite, there exists a finite subset F of Φ for which $\bigcup_{\varphi \in \Phi} \varphi^* = \bigcup_{f \in F} f^*$. We show $\Phi = \bigcup_{f \in F} f\alpha_\Phi$. Let $\varphi \in \Phi$. Select $t \in \varphi^*$. Now $t \in f^*$ for some $f \in F$. Hence $A\varphi \subset At\beta^2$ $(A \in \mathfrak{A})$ and $Af \subset At\beta^2$ $(A \in \mathfrak{A})$. If $x \in X$, then $x \in A$ for some $A \in \mathfrak{A}$ whence $x\varphi \in At\beta^2$, $xf \in At\beta^2$ and $(xf, x\varphi) = (xf, At)\,(At, x\varphi) \in \beta^4 \subset \alpha$. Thus $(xf, x\varphi) \in \alpha$ $(x \in X)$ and $\varphi \in f\alpha_\Phi$. The proof is completed.

11.07. DEFINITION. Let X be a topological space, let Y be a uniform space and let Φ be a set of mappings of X into Y. If $x \in X$, then Φ is said to be *equicontinuous at x* provided that if β is an index of Y, then there exists a neighborhood U of x such that $\varphi \in \Phi$ implies $U\varphi \subset x\varphi\beta$. The set Φ is said to be *equicontinuous* provided that Φ is equicontinuous at every point of X.

11.08. DEFINITION. Let X, Y be uniform spaces and let Φ be a set of mappings of X into Y. The set Φ is said to be *uniformly equicontinuous* provided that if β is an index of Y, then there exists an index α of X such that $x \in X$ and $\varphi \in \Phi$ implies $x\alpha\varphi \subset x\varphi\beta$.

11.09. Remark. Let X and Y be uniform spaces and let Φ be a set of mappings of X into Y. Then:

(1) If Φ is uniformly equicontinuous, then Φ is equicontinuous.

(2) If X is compact and if Φ is equicontinuous, then Φ is uniformly equicontinuous.

11.10. Remark. Let X be a {topological}{uniform} space, let Y be a uniform space, let Φ be the set of all mappings of X into Y, let Φ be provided with its space-index topology and let Ψ be {an equicontinuous} {a uniformly equicontinuous} set of mappings of X into Y. Then Ψ is {equicontinuous} {uniformly equicontinuous}.

11.11. Definition. Let X be a set, let Y be a uniform space and let Φ be a set of mappings of X into Y. The set Φ is said to be *bounded* provided that $\bigcup_{\varphi \in \Phi} X\varphi$ is a totally bounded subset of Y.

11.12. Theorem. *Let X and Y be uniform spaces and let Φ be a set of uniformly continuous mappings of X into Y. Then:*

(1) *If Φ is totally bounded in its space-index uniformity, then Φ is uniformly equicontinuous.*

(2) *If X is totally bounded, then Φ is totally bounded in its space-index uniformity if and only if Φ is uniformly equicontinuous and bounded.*

Proof. (1) Let β be an index of Y. Choose a symmetric index γ of Y for which $\gamma^3 \subset \beta$. Since Φ is totally bounded in its space-index uniformity, there exists a finite subset F of Φ such that $\varphi \in \Phi$ implies $(xf, x\varphi) \in \gamma$ $(x \in X)$ for some $f \in F$. Select an index α of X such that $(x_1, x_2) \in \alpha$ and $f \in F$ implies $(x_1 f, x_2 f) \in \gamma$. We show that $(x_1, x_2) \in \alpha$ and $\varphi \in \Phi$ implies $(x_1\varphi, x_2\varphi) \in \beta$. Let $(x_1, x_2) \in \alpha$ and let $\varphi \in \Phi$. Choose $f \in F$ such that $(xf, x\varphi) \in \gamma$ $(x \in X)$. Then $(x_1\varphi, x_2\varphi) = (x_1\varphi, x_1 f)(x_1 f, x_2 f)(x_2 f, x_2\varphi) \in \gamma^3 \subset \beta$. This proves (1).

(2) Assume that X is totally bounded and that Φ is uniformly equicontinuous and bounded. We show that Φ is totally bounded in its space-index uniformity. It is enough by 11.06 to show that for each index β of Y there exists a finite partition $\mathfrak{A}$ of X such that $A \in \mathfrak{A}$ and $\varphi \in \Phi$ implies $A\varphi \times A\varphi \subset \beta$. Let β be an index of Y. Choose an index α of X such that $(x_1, x_2) \in \alpha$ and $\varphi \in \Phi$ implies $(x_1\varphi, x_2\varphi) \in \beta$. Since X is totally bounded, there exists a finite partition $\mathfrak{A}$ of X such that $A \in \mathfrak{A}$ implies $A \times A \subset \alpha$. Hence $A \in \mathfrak{A}$ and $\varphi \in \Phi$ implies $A\varphi \times A\varphi \subset \beta$.

Now assume that X is totally bounded and that Φ is totally bounded in its space-index uniformity. By (1), Φ is uniformly equicontinuous. We show that Φ is bounded, that is, $\bigcup_{\varphi \in \Phi} X\varphi$ is totally bounded. Let β be an index of Y. Choose an index γ of Y for which $\gamma^2 \subset \beta$. There exists a finite subset F of Φ such that $\varphi \in \Phi$ implies $(xf, x\varphi) \in \gamma$ $(x \in X)$ for some $f \in F$. Since each $f \in F$ is uniformly continuous, $\bigcup_{f \in F} Xf$ is totally bounded. Hence there exists a finite subset E of Y such that $\bigcup_{f \in F} Xf \subset E\gamma$. Since $\varphi \in \Phi$ implies $X\varphi \subset Xf\gamma$, it follows that $\bigcup_{\varphi \in \Phi} X\varphi \subset \bigcup_{f \in F} Xf\gamma \subset E\gamma^2 \subset E\beta$. The proof is completed.

11.13. THEOREM. *Let X and Y be compact uniform spaces, let Φ be the set of all continuous mappings of X into Y, let Φ be provided with its space-index topology and let $\Psi \subset \Phi$. Then Ψ is compact if and only if Ψ is equicontinuous.*

PROOF. Use 11.05, 11.09 and 11.12.

11.14. THEOREM. *Let X be a uniform space and let Φ be a semigroup of uniformly continuous mappings of X into X. Then the semigroup multiplication of Φ is continuous in the space-index topology of Φ.*

PROOF. Let φ_0, $\psi_0 \in \Phi$ and let γ be an index of X. Choose an index β of X such that $\beta^2 \subset \gamma$. There exists an index of α of X such that $(x, y) \in \alpha$ implies $(x\psi_0, y\psi_0) \in \beta$. We show $(\varphi_0\alpha_\Phi)(\psi_0\beta_\Phi) \subset (\varphi_0\psi_0)\gamma_\Phi$. Let $\varphi \in \varphi_0\alpha_\Phi$ and let $\psi \in \psi_0\beta_\Phi$. If $x \in X$, then $(x\varphi_0, x\varphi) \in \alpha$, $(x\varphi_0\psi_0, x\varphi\psi_0) \in \beta$, $(x\varphi\psi_0, x\varphi\psi) \in \beta$ and $(x\varphi_0\psi_0, x\varphi\psi) \in \beta^2 \subset \gamma$. Hence $\varphi\psi \in (\varphi_0\psi_0)\gamma_\Phi$. The proof is completed.

11.15. DEFINITION. Let X be a set. A *permutation of X* is defined to be a one-to-one mapping of X onto X.

11.16. DEFINITION. Let X be a uniform space with uniformity I and let Φ be a set of permutations of X. For $\alpha \subset X \times X$ define

$$\alpha_\Phi = [(\varphi, \psi) \mid \varphi, \psi \in \Phi \quad \text{and} \quad (x\varphi, x\psi) \in \alpha\ (x \in X)],$$
$$\alpha_\Phi^* = [(\varphi, \psi) \mid \varphi, \psi \in \Phi \quad \text{and} \quad (x\varphi^{-1}, x\psi^{-1}) \in \alpha\ (x \in X)],$$
$$\hat{\alpha}_\Phi = \alpha_\Phi \cap \alpha_\Phi^* .$$

Define

$$\mathfrak{V} = [\alpha_\Phi \mid \alpha \in I],$$
$$\mathfrak{V}^* = [\alpha_\Phi^* \mid \alpha \in I],$$
$$\hat{\mathfrak{V}} = [\hat{\alpha}_\Phi \mid \alpha \in I].$$

It is readily verified that $\mathfrak{V}$, $\mathfrak{V}^*$, $\hat{\mathfrak{V}}$ are uniformity-bases of Φ. The uniformities $\mathfrak{U}$, $\mathfrak{U}^*$, $\hat{\mathfrak{U}}$ of Φ generated by $\mathfrak{V}$, $\mathfrak{V}^*$, $\hat{\mathfrak{V}}$ are called the *space-index*, the *inverse space-index*, the *bilateral space-index* uniformities of Φ. The topologies $\mathfrak{J}$, $\mathfrak{J}^*$, $\hat{\mathfrak{J}}$ of Φ induced by $\mathfrak{U}$, $\mathfrak{U}^*$, $\hat{\mathfrak{U}}$ are called the *space-index*, the *inverse space-index*, the *bilateral space-index* topologies of Φ. It is clear that:

(1) If Φ is a group, then $\{\mathfrak{U}^*\}\{\mathfrak{J}^*\}$ is the image of $\{\mathfrak{U}\}\{\mathfrak{J}\}$ under the group inversion of Φ.

(2) $\hat{\mathfrak{U}} = \mathfrak{U} \vee \mathfrak{U}^*$, $\hat{\mathfrak{J}} = \mathfrak{J} \vee \mathfrak{J}^*$.

11.17. REMARK. Let G be a group, let $\{\mathfrak{J}\}\{\mathfrak{U}\}$ be a $\{$topology$\}\{$uniformity$\}$ of G, let $\{\mathfrak{J}^*\}\{\mathfrak{U}^*\}$ be the image of $\{\mathfrak{J}\}\{\mathfrak{U}\}$ under the group inversion of G and let $\{\hat{\mathfrak{J}} = \mathfrak{J} \vee \mathfrak{J}^*\}\{\hat{\mathfrak{U}} = \mathfrak{U} \vee \mathfrak{U}^*\}$. Then:

(1) The group inversion of G is a homeomorphism of $\{(G, \mathfrak{J})$ onto $(G, \mathfrak{J}^*)\}$ $\{(G, \mathfrak{J}^*)$ onto $(G, \mathfrak{J})\}\{(G, \hat{\mathfrak{J}})$ onto $(G, \hat{\mathfrak{J}})\}$.

(2) The group inversion of G is a unimorphism of $\{(G, \mathfrak{U})$ onto $(G, \mathfrak{U}^*)\}$ $\{(G, \mathfrak{U}^*)$ onto $(G, \mathfrak{U})\}\{(G, \hat{\mathfrak{U}})$ onto $(G, \hat{\mathfrak{U}})\}$.

(3) The following statements are pairwise equivalent:

(I) The group inversion of G is $\mathfrak{J}$-continuous and hence $\mathfrak{J}$-homeomorphic.

(II) The group inversion of G is $\mathfrak{J}^*$-continuous and hence $\mathfrak{J}^*$-homeomorphic.

(III) $\mathfrak{J} = \mathfrak{J}^*$ and hence $\mathfrak{J} = \mathfrak{J}^* = \hat{\mathfrak{J}}$.

(4) The following statements are pairwise equivalent:

(I) The group inversion of G is $\mathfrak{U}$-uniformly continuous and hence $\mathfrak{U}$-unimorphic.

(II) The group inversion of G is $\mathfrak{U}^*$-uniformly continuous and hence $\mathfrak{U}$-unimorphic.

(III) $\mathfrak{U} = \mathfrak{U}^*$ and hence $\mathfrak{U} = \mathfrak{U}^* = \hat{\mathfrak{U}}$.

(5) If the group multiplication of G is $\mathfrak{J}$-continuous, then $(G, \hat{\mathfrak{J}})$ is a topological group.

(6) If $\mathfrak{U}$ induces $\mathfrak{J}$, then $\{\mathfrak{U}^*$ induces $\mathfrak{J}^*\}\{\hat{\mathfrak{U}}$ induces $\hat{\mathfrak{J}}\}$.

(7) If G provided with some topology is a topological group and if $\mathfrak{U}_L$, $\mathfrak{U}_R$, $\mathfrak{U}_B$ are the left, right, bilateral uniformities of G, then:

(I) $\mathfrak{U} \subset \mathfrak{U}_L$ is equivalent to $\mathfrak{U}^* \subset \mathfrak{U}_R$, and either condition implies $\hat{\mathfrak{U}} \subset \mathfrak{U}_B$.

(II) $\mathfrak{U} \supset \mathfrak{U}_L$ is equivalent to $\mathfrak{U}^* \supset \mathfrak{U}_R$, and either condition implies $\hat{\mathfrak{U}} \supset \mathfrak{U}_B$.

(III) $\mathfrak{U} = \mathfrak{U}_L$ is equivalent to $\mathfrak{U}^* = \mathfrak{U}_R$, and either condition implies $\hat{\mathfrak{U}} = \mathfrak{U}_B$.

11.18. Theorem. *Let X be a uniform space and let Φ be a group of unimorphisms of X onto X. Then:*

(1) *The space-index, the inverse space-index, the bilateral space-index topologies of Φ all coincide; call this topology $\mathfrak{J}$.*

(2) *$(\Phi, \mathfrak{J})$ is a topological group.*

(3) *The $\{$left$\}\{$right$\}\{$bilateral$\}$ uniformity of $(\Phi, \mathfrak{J})$ coincides with the $\{$space-index$\}\{$inverse space-index$\}\{$bilateral space-index$\}$ uniformity of Φ.*

(4) *$(\Phi, \mathfrak{J})$ is a topological homeomorphism group of X.*

Proof. (1) By 11.17 (3) it is enough to show that the group inversion of Φ is continuous with respect to the space-index topology of Φ. Let $\varphi \in \Phi$ and let α be a symmetric index of X. There exists an index β of X such that $(x, y) \in \beta$ implies $(x\varphi^{-1}, y\varphi^{-1}) \in \alpha$. We show $(\varphi\beta_\Phi)^{-1} \subset \varphi^{-1}\alpha_\Phi$. Let $\psi \in \varphi\beta_\Phi$. Then $(x\varphi, x\psi) \in \beta$ $(x \in X)$, $(x, x\psi\varphi^{-1}) = (x\varphi\varphi^{-1}, x\psi\varphi^{-1}) \in \alpha$ $(x \in X)$, $(x\psi^{-1}, x\varphi^{-1}) = (x\psi^{-1}, x\psi^{-1}\psi\varphi^{-1}) \in \alpha$ $(x \in X)$, $(x\varphi^{-1}, x\psi^{-1}) \in \alpha$ $(x \in X)$ and $\psi^{-1} \in \varphi^{-1}\alpha_\Phi$. This proves (1).

(2) Use 11.14 and 11.17 (5).

(3) Let $\varphi, \psi \in \Phi$, let θ be the identity mapping of X and let α be an index of X. The following statements are readily verified:

$$(\varphi, \psi) \in \alpha_\Phi \Leftrightarrow \varphi^{-1}\psi \in \theta\alpha_\Phi ,$$

$$(\varphi, \psi) \in \alpha_\Phi^* \Leftrightarrow \varphi\psi^{-1} \in \theta\alpha_\Phi ,$$

$$(\varphi, \psi) \in \hat{\alpha}_\Phi \Leftrightarrow \varphi^{-1}\psi \in \theta\alpha_\Phi \& \varphi\psi^{-1} \in \theta\alpha_\Phi .$$

The conclusion follows. Cf. also 11.17 (7).

(4) Use 11.02.

11.19. THEOREM. *Let X be a complete separated uniform space and let Φ be the group of all unimorphisms of X onto X. Then Φ is complete in its bilateral space-index uniformity.*

PROOF. Let $\mathfrak{U}$ be the bilateral space-index uniformity of Φ and let $\mathfrak{F}$ be a $\mathfrak{U}$-cauchy filter on Φ. Let Ψ be the semigroup of all uniformly continuous mappings of X into X and let $\mathfrak{V}$ be the space-index uniformity of Ψ. By 11.05 there exist $\varphi, \psi \in \Psi$ such that $\mathfrak{F} \to \varphi$ in $\mathfrak{V}$ and $\mathfrak{F}^{-1} \to \psi$ in $\mathfrak{V}$. By 11.14. $\mathfrak{F}\mathfrak{F}^{-1} \to \varphi\psi$ in $\mathfrak{V}$ and $\mathfrak{F}^{-1}\mathfrak{F} \to \psi\varphi$ in $\mathfrak{V}$. Let θ be the identity mapping of X. Since $\theta \in \bigcap \mathfrak{F}\mathfrak{F}^{-1}$ and $\theta \in \bigcap \mathfrak{F}^{-1}\mathfrak{F}$, we have $\varphi\psi = \theta = \psi\varphi$. Hence $\varphi, \psi \in \Phi$, $\varphi^{-1} = \psi$ and $\mathfrak{F} \to \varphi$ in $\mathfrak{U}$. The proof is completed.

11.20. THEOREM. ***Let X be a complete separated uniform space and let Φ be the group of all unimorphisms of X onto X. Then Φ is a bilaterally complete topological group in its space-index topology.***

PROOF. Use 11.18 (2) and 11.19.

11.21. THEOREM. *Let X be a totally bounded uniform space and let Φ be a symmetric set of unimorphisms of X onto X. Then Φ is totally bounded in its bilateral space-index uniformity if and only if Φ is uniformly equicontinuous.*

PROOF. Use 11.18 and 11.12 (2).

11.22. THEOREM. *Let X be a compact uniform space, let Φ be the group of all homeomorphisms of X onto X, let Φ be provided with its space-index topology and let Ψ be a symmetric subset of Φ. Then $\overline{\Psi}$ is compact if and only if Ψ is equicontinuous.*

PROOF. Use 11.18 and 11.21.

11.23. REMARK. Let X and Y be sets, let Z be a uniform space and let $\pi : X \times Y \to Z$. Write $(x, y)\pi = xy$ $(x \in X, y \in Y)$. For $x \in X$ define $\pi_x : Y \to Z$ by $y\pi_x = xy$ $(y \in Y)$. For $y \in Y$ define $\pi^y : X \to Z$ by $x\pi^y = xy$ $(x \in X)$. Then:

(1) If X is a totally bounded uniform space and if $[\pi^y \mid y \in Y]$ is uniformly equicontinuous, then $[\pi_x \mid x \in X]$ is totally bounded in its space-index uniformity.

(2) If $[\pi_x \mid x \in X]$ is totally bounded in its space-index uniformity, then for each index α of Z there exists a finite partition $\mathfrak{A}$ of X such that $A \in \mathfrak{A}$ and $y \in Y$ implies $Ay \times Ay \subset \alpha$.

11.24. DEFINITION. Let X be a topological space, let $\mathfrak{A}$ be the class of all compact subsets of X, let Y be a uniform space with uniformity I and let Φ be a set of mappings of X into Y. For $A \subset X$ and $\alpha \subset Y \times Y$ define

$$(A, \alpha)_\Phi = [(\varphi, \psi) \mid \varphi, \psi \in \Phi; (x\varphi, x\psi) \in \alpha \ (x \in A)].$$

Define

$$\mathfrak{V} = [(A, \alpha)_\Phi \mid A \in \mathfrak{A}, \alpha \in I].$$

It is readily verified that $\mathcal{V}$ is a uniformity-base of Φ. The uniformity $\mathcal{U}$ of Φ generated by $\mathcal{V}$ is called the *compact-index* uniformity of Φ. The topology of Φ induced by $\mathcal{U}$ is called the *compact-index* topology of Φ.

When the domain space X is discrete, the term *compact-index* is replaced by the term *point-index*.

11.25. REMARK. Let X be a compact topological space, let Y be a uniform space and let Φ be a set of mappings of X into Y. Then the space-index uniformity of Φ coincides with the compact-index uniformity of Φ.

11.26. THEOREM. *Let X be a locally compact topological space, let Y be a uniform space, let Φ be a set of continuous mappings of X into Y, let Φ be provided with its compact-index topology and let $\pi : X \times \Phi \to Y$ be defined by $(x, \varphi)\pi = x\varphi$ $(x \in X, \varphi \in \Phi)$. Then π is continuous.*

PROOF. Let $x \in X$, let $\varphi \in \Phi$ and let α be an index of Y. Choose an index β of Y such that $\beta^2 \subset \alpha$. There exists a compact neighborhood U of x such that $U\varphi \subset x\varphi\beta$. It follows that $(U \times \varphi(U, \beta)_\Phi)\pi \subset x\varphi\alpha$. The proof is completed.

11.27. THEOREM. *Let X be a topological space, let Y be a complete uniform space and let Φ be the set of all mappings of X into Y. Then Φ is complete in its compact-index uniformity.*

PROOF. Let Φ be provided with its compact-index uniformity. Let $\mathfrak{F}$ be a cauchy filter on Φ. For each $x \in X$, $[xF \mid F \in \mathfrak{F}]$ is a cauchy filter-base on Y. Define $\varphi : X \to Y$ by $x\varphi \in \bigcap_{F \in \mathfrak{F}} \overline{xF}$ $(x \in X)$. We show $\mathfrak{F} \to \varphi$. Let A be a compact subset of X and let α be a closed index of Y. There exists $F \in \mathfrak{F}$ such that $F \times F \subset (A, \alpha)_\Phi$. Then $xF \times xF \subset \alpha$ $(x \in A)$, $\overline{xF} \times xF \subset \alpha$ $(x \in A)$, $[x\varphi] \times xF \subset \alpha$ $(x \in A)$, $[\varphi] \times F \subset (A, \alpha)_\Phi$ and $F \subset \varphi(A, \alpha)_\Phi$. The proof is completed.

11.28. THEOREM. *Let X be a locally compact topological space, let Y be a uniform space, let Φ be the set of all mappings of X into Y, let Φ be provided with its compact-index topology and let Ψ be the set of all continuous mappings of X into Y. Then Ψ is a closed subset of Φ.*

PROOF. Let $\varphi \in \overline{\Psi}$. We show φ is continuous. Let $x_0 \in X$ and let α be an index of Y. Choose a symmetric index β of Y such that $\beta^3 \subset \alpha$ and choose a compact neighborhood V of x_0. There exists $\psi \in \Psi \cap \varphi(V, \beta)_\Phi$. Select a neighborhood U of x_0 such that $U \subset V$ and $U\psi \subset x_0\psi\beta$. If $x \in U$, then $(x_0\varphi, x_0\psi) \in \beta$, $(x_0\psi, x\psi) \in \beta$, $(x\psi, x\varphi) \in \beta$, $(x_0\varphi, x\varphi) \in \beta^3 \subset \alpha$ and $x\varphi \in x_0\varphi\alpha$. The proof is completed.

11.29. THEOREM. *Let X be a locally compact topological space, let Y be a complete uniform space and let Φ be the set of all continuous mappings of X into Y. Then Φ is complete in its compact-index uniformity.*

PROOF. Use 11.27 and 11.28.

11.30. Definition. Let X be a topological space, let Y be a uniform space and let Φ be a set of mappings of X into Y. The set Φ is said to be *locally bounded* provided that if $x \in X$, then there exists a neighborhood U of x such that $\bigcup_{\varphi \in \Phi} U\varphi$ is a totally bounded subset of Y.

11.31. Theorem. *Let X be a locally compact T_2-space, let Y be a uniform space and let Φ be a set of continuous mappings of X into Y. Then Φ is totally bounded in its compact-index uniformity if and only if Φ is equicontinuous and locally bounded.*

Proof. Apply 11.12 to the restrictions of Φ to the compact subsets of X.

11.32. Theorem. *Let X be a locally compact T_2-space, let Y be a complete uniform space, let Φ be the set of all continuous mappings of X into Y, let Φ be provided with its compact-index topology and let $\Psi \subset \Phi$. Then $\overline{\Psi}$ is compact if and only if Ψ is equicontinuous and locally bounded.*

Proof. Use 11.29 and 11.31.

11.33. Theorem. *Let X be a locally compact uniform space and let Φ be a semigroup of continuous mappings of X into X. Then the semigroup multiplication of Φ is continuous in the compact-index topology of Φ.*

Proof. Let $\varphi_0, \psi_0 \in \Phi$, let A be a compact subset of X and let α be an index of X. Choose a compact subset C of X and an index γ of X such that $A\varphi_0\gamma \subset C$. Then choose an index β of X such that $\beta \subset \gamma$ and such that $(x_1, x_2) \in \beta \cap (C \times C)$ implies $(x_1\psi_0, x_2\psi_0) \in \gamma$. Define $B = A \cup C$. Let $(\varphi_0, \varphi) \in (B, \beta)_\Phi$ and let $(\psi_0, \psi) \in (C, \gamma)_\Phi$. We show $(\varphi_0\psi_0, \varphi\psi) \in (A, \alpha)_\Phi$. Let $x \in A$. Then $x \in B$ and $(x\varphi_0, x\varphi) \in \beta$. Since $(a\varphi_0, a\varphi) \in \beta \subset \gamma$ $(a \in A)$, we have $A\varphi \subset A\varphi_0\gamma \subset C$ and $x\varphi_0, x\varphi \in C$. Hence $(x\varphi_0, x\varphi) \in \beta \cap (C \times C)$. We conclude that $(x\varphi_0\psi_0, x\varphi\psi_0) \in \gamma$, $(x\varphi\psi_0, x\varphi\psi) \in \gamma$ and $(x\varphi_0\psi_0, x\varphi\psi) \in \gamma^2 \subset \alpha$. This shows that $(\varphi_0\psi_0, \varphi\psi) \in (A, \alpha)_\Phi$. The proof is completed.

11.34. Definition. Let X be a uniform space with uniformity I, let $\mathfrak{A}$ be the class of all compact subsets of X and let Φ be a set of permutations of X. For $A \subset X$ and $\alpha \subset X \times X$ define

$$(A, \alpha)_\Phi = [(\varphi, \psi) \mid \varphi, \psi \in \Phi \quad \text{and} \quad (x\varphi, x\psi) \in \alpha\ (x \in A)],$$

$$(A, \alpha)_\Phi^* = [(\varphi, \psi) \mid \varphi, \psi \in \Phi \quad \text{and} \quad (x\varphi^{-1}, x\psi^{-1}) \in \alpha\ (x \in A)],$$

$$(A, \alpha)_\Phi^\wedge = (A, \alpha)_\Phi \cap (A, \alpha)_\Phi^* .$$

Define

$$\mathcal{V} = [(A, \alpha)_\Phi \mid A \in \mathfrak{A} \quad \text{and} \quad \alpha \in I],$$

$$\mathcal{V}^* = [(A, \alpha)_\Phi^* \mid A \in \mathfrak{A} \quad \text{and} \quad \alpha \in I],$$

$$\hat{\mathcal{V}} = [(A, \alpha)_\Phi^\wedge \mid A \in \mathfrak{A} \quad \text{and} \quad \alpha \in I].$$

It is readily verified that $\mathcal{V}$, $\mathcal{V}^*$, $\hat{\mathcal{V}}$ are uniformity-bases of Φ. The uniformities

$\mathfrak{U}$, $\mathfrak{U}^*$, $\hat{\mathfrak{U}}$ of Φ generated by $\mathfrak{V}$, $\mathfrak{V}^*$, $\hat{\mathfrak{V}}$ are called the *compact-index*, the *inverse compact-index*, the *bilateral compact-index* uniformities of Φ. The topologies $\mathfrak{J}$, $\mathfrak{J}^*$, $\hat{\mathfrak{J}}$ of Φ induced by $\mathfrak{U}$, $\mathfrak{U}^*$, $\hat{\mathfrak{U}}$ are called the *compact-index*, the *inverse compact-index*, the *bilateral compact-index* topologies of Φ. It is clear that:

(1) If Φ is a group, then $\{\mathfrak{U}^*\}\{\mathfrak{J}^*\}$ is the image of $\{\mathfrak{U}\}\{\mathfrak{J}\}$ under the group inversion of Φ.

(2) $\hat{\mathfrak{U}} = \mathfrak{U} \vee \mathfrak{U}^*$, $\hat{\mathfrak{J}} = \mathfrak{J} \vee \mathfrak{J}^*$.

11.35. Remark. Let X be a compact uniform space and let Φ be a group of permutations of X. Then the {space-index}{inverse space-index}{bilateral space-index} uniformity of Φ coincides with the {compact-index}{inverse compact-index}{bilateral compact-index} uniformity of Φ.

11.36. Theorem. *Let X be a locally compact uniform space, let Φ be a group of homeomorphisms of X onto X and let Φ be provided with its bilateral compact-index topology. Then:*

(1) *Φ is a topological group.*

(2) *Φ is a topological homeomorphism group of X.*

Proof. Use 11.33.

11.37. Theorem. *Let X be a locally compact complete separated uniform space and let Φ be the group of all homeomorphisms of X onto X. Then Φ is complete in its bilateral compact-index uniformity.*

Proof. Let $\mathfrak{U}$ be the bilateral compact-index uniformity of Φ and let $\mathfrak{F}$ be a $\mathfrak{U}$-cauchy filter on Φ. Let Ψ be the semigroup of all continuous mappings of X into X and let $\mathfrak{V}$ be the compact-index uniformity of Ψ. By 11.29 there exist $\varphi, \psi \in \Psi$ such that $\mathfrak{F} \to \varphi$ in $\mathfrak{V}$ and $\mathfrak{F}^{-1} \to \psi$ in $\mathfrak{V}$. By 11.33, $\mathfrak{F}\mathfrak{F}^{-1} \to \varphi\psi$ in $\mathfrak{V}$ and $\mathfrak{F}^{-1}\mathfrak{F} \to \psi\varphi$ in $\mathfrak{V}$. Let θ be the identity mapping of X. Since $\theta \in \bigcap \mathfrak{F}\mathfrak{F}^{-1}$ and $\theta \in \bigcap \mathfrak{F}^{-1}\mathfrak{F}$, we have $\varphi\psi = \theta = \psi\varphi$. Hence, $\varphi, \psi \in \Phi$, $\varphi^{-1} = \psi$ and $\mathfrak{F} \to \varphi$ in $\mathfrak{U}$. The proof is completed.

11.38. Theorem. ***Let X be a locally compact complete separated uniform space and let Φ be the group of all homeomorphisms of X onto X. Then Φ is a bilaterally complete topological group in its bilateral compact-index topology.***

Proof. Let Φ be provided with its bilateral compact-index topology. By 11.36, Φ is a topological group. Let $\mathfrak{F}$ be a filter on Φ which is a cauchy filter in the bilateral uniformity of Φ. By 11.37 it is enough to show that $\mathfrak{F}$ is a cauchy filter in the bilateral compact-index uniformity of Φ.

We first show that if A is a compact subset of X, then there exists $F \in \mathfrak{F}$ such that $AF \cup AF^{-1}$ is a conditionally compact subset of X. Let A be a compact subset of X. Choose an index α of X such that $A\alpha$ is a conditionally compact subset of X. Then there exists $F \in \mathfrak{F}$ such that $\varphi, \psi \in F$ implies $(x, x\varphi^{-1}\psi) \in \alpha$ $(x \in A)$ and $(x, x\varphi\psi^{-1}) \in \alpha$ $(x \in A)$. Select $\varphi_0 \in F$. Then $\varphi \in F$ implies $x\varphi^{-1}\varphi_0 \in x\alpha$ $(x \in A)$, $x\varphi\varphi_0^{-1} \in x\alpha$ $(x \in A)$, $A\varphi^{-1}\varphi_0 \subset A\alpha$, $A\varphi\varphi_0^{-1} \subset A\alpha$, $A\varphi^{-1} \subset A\alpha\varphi_0^{-1}$, and $A\varphi \subset A\alpha\varphi_0$. Hence $AF \cup AF^{-1} \subset A\alpha\varphi_0^{-1} \cup A\alpha\varphi_0$ and $AF \cup AF^{-1}$ is conditionally compact.

Let A be a compact subset of X and let α be an index of X. Choose a compact subset B of X such that $AF_1 \cup AF_1^{-1} \subset B$ for some $F_1 \in \mathfrak{F}$. There exists $F_2 \in \mathfrak{F}$ such that $\varphi, \psi \in F_2$ implies $(y, y\varphi^{-1}\psi) \in \alpha$ $(y \in B)$ and $(y, y\varphi\psi^{-1}) \in \alpha$ $(y \in B)$. Define $F = F_1 \cap F_2 \in \mathfrak{F}$. Then $\varphi, \psi \in F$ implies $(x\varphi, x\psi) \in \alpha$ $(x \in A)$ and $(x\varphi^{-1}, x\psi^{-1}) \in \alpha$ $(x \in A)$.

Hence, $\mathfrak{F}$ is a cauchy filter in the bilateral compact-index uniformity of Φ. The proof is completed.

11.39. Theorem. *Let X be a locally connected locally compact uniform space and let Φ be a group of homeomorphisms of X. Then the compact-index, the inverse compact-index, and the bilateral compact-index topologies of Φ all coincide.*

Proof. It is enough to show that the group inversion of Φ is continuous in the compact-index topology of Φ. Let A be a compact connected subset of X such that int $A \neq \emptyset$, let α be an index of X and let $\varphi \in \Phi$. It is enough to show there exist a compact subset B of X and an index β of X such that $(\varphi, \psi) \in (B, \beta)_\Phi$ implies $(\varphi^{-1}, \psi^{-1}) \in (A, \alpha)_\Phi$.

We first show there exist a compact subset C of X and an index γ of X such that $(\varphi, \psi) \in (C, \gamma)_\Phi$ implies $A\psi^{-1} \subset C$. Choose a compact neighborhood U of A and then choose a symmetric index γ of X such that $A\gamma \subset$ int U and $a\gamma \subset A$ for some $a \in A$. Define $C = U\varphi^{-1}$. Let $(\varphi, \psi) \in (C, \gamma)_\Phi$. We show $A\psi^{-1} \subset C$. Suppose $A\psi^{-1} \not\subset C$. Then $A \not\subset C\psi$. Since $a\varphi^{-1} \in C$ and $a\varphi^{-1}\psi \in a\varphi^{-1}\varphi\gamma = a\gamma \subset A$, we have $A \cap C\psi \neq \emptyset$. Since A is connected, there exists $x \in A \cap$ bdy $(C\psi)$. Define $y = x\psi^{-1}$. It follows that $y\varphi \in$ (bdy $C)\varphi =$ bdy U and $y\psi = x \in A$. Since $y \in C$, we have $(y\varphi, y\psi) \in \gamma$ and $y\varphi \in y\psi\gamma \subset A\gamma \subset$ int U. We now have $y\varphi \in$ bdy U and $y\varphi \in$ int U. This is a contradiction. Hence $A\psi^{-1} \subset C$.

Select a compact subset C of X and an index γ of X such that $(\varphi, \psi) \in (C, \gamma)_\Phi$ implies $A\psi^{-1} \subset C$. Define $B = C \cup C\varphi$. Choose a symmetric index β of X such that $\beta \subset \gamma$ and such that $(x_1, x_2) \in \beta \cap (B \times B)$ implies $(x_1\varphi^{-1}, x_2\varphi^{-1}) \in \alpha$. Let $(\varphi, \psi) \in (B, \beta)_\Phi$. We show $(\varphi^{-1}, \psi^{-1}) \in (A, \alpha)_\Phi$. Let $x \in A$. Since $(\varphi, \varphi) \in (C, \gamma)_\Phi$, we have $A\varphi^{-1} \subset C$, $x \in A \subset C\varphi \subset B$ and $x \in B$. Since $(\varphi, \psi) \in (C, \gamma)_\Phi$, we have $A\psi^{-1} \subset C$, $x\psi^{-1} \in C$, $x\psi^{-1}\varphi \in C\varphi \subset B$ and $x\psi^{-1}\varphi \in B$. Since $x\psi^{-1} \in B$, and $(\varphi, \psi) \in (B, \beta)_\Phi$, we have $(x\psi^{-1}\varphi, x) \in \beta$ and $(x, x\psi^{-1}\varphi) \in \beta$. Thus $(x, x\psi^{-1}\varphi) \in \beta \cap (B \times B)$. It follows that $(x\varphi^{-1}, x\psi^{-1}) \in \alpha$. Hence $(\varphi^{-1}, \psi^{-1}) \in (A, \alpha)_\Phi$. The proof is completed.

11.40. Definition. Let X and Y be topological spaces, let $\mathfrak{A}$ be the class of all compact subsets of X, let $\mathcal{E}$ be the class of all open subsets of Y and let Φ be a set of mappings of X into Y. For $A \subset X$ and $E \subset Y$ define

$$(A, E)_\Phi = [\varphi \mid \varphi \in \Phi \quad \text{and} \quad A\varphi \subset E].$$

Define

$$\mathcal{S} = [(A, E)_\Phi \mid A \in \mathfrak{A} \quad \text{and} \quad E \in \mathcal{E}].$$

The topology $\mathfrak{I}$ of Φ generated by $\mathcal{S}$ is called the *compact-open* topology of Φ.

11.41. Theorem. *Let X be a topological space, let Y be a uniform space and let Φ be a set of continuous mappings of X into Y. Then the compact-index topology of Φ coincides with the compact-open topology of Φ.*

Proof. Let $\mathfrak{I}_1$ be the compact-index topology of Φ and let $\mathfrak{I}_2$ be the compact-open topology of Φ.

We show $\mathfrak{I}_1 \subset \mathfrak{I}_2$. Let A be a compact subset of X, let α be an index of Y and let $\varphi \in \Phi$. It is enough to show that $U \subset \varphi(A, \alpha)_\Phi$ for some $\mathfrak{I}_2$-neighborhood U of φ. Since φ is continuous on the compact set A, there exist a finite family $(A_\iota \mid \iota \in I)$ of closed subsets of A and a family $(E_\iota \mid \iota \in I)$ of open subsets of Y such that $A = \bigcup_{\iota \in I} A_\iota$, $A_\iota\varphi \subset E_\iota$ $(\iota \in I)$ and $E_\iota \times E_\iota \subset \alpha$ $(\iota \in I)$. Define $U = \bigcap_{\iota \in I} (A_\iota, E_\iota)_\Phi$. Now U is a $\mathfrak{I}_2$-neighborhood of φ. If $\psi \in U$ and if $x \in A$, then $x \in A_\iota$ for some $\iota \in I$ whence $(x\varphi, x\psi) \in A_\iota\varphi \times A_\iota\psi \subset E_\iota \times E_\iota \subset \alpha$. Therefore $\psi \in U$ implies $(x\varphi, x\psi) \in \alpha$ $(x \in A)$ and $\psi \in \varphi(A, \alpha)_\Phi$. Thus $U \subset \varphi(A, \alpha)_\Phi$. This shows $\mathfrak{I}_1 \subset \mathfrak{I}_2$.

We show $\mathfrak{I}_2 \subset \mathfrak{I}_1$. Let A be a compact subset of X, let E be an open subset of Y and let $\varphi \in (A, E)_\Phi$. It is enough to show that $U \subset (A, E)_\Phi$ for some $\mathfrak{I}_1$-neighborhood U of φ. Since $A\varphi \subset E$ and $A\varphi$ is compact, there exists an index α of Y such that $A\varphi\alpha \subset E$. Define $U = \varphi(A, \alpha)_\Phi$. Now U is a $\mathfrak{I}_1$-neighborhood of φ. If $\psi \in U$, then $(x\varphi, x\psi) \in \alpha$ $(x \in A)$, $x\psi \in x\varphi\alpha \subset E$ $(x \in A)$, $A\psi \subset E$ and $\psi \in (A, E)_\Phi$. Thus $U \subset (A, E)_\Phi$. This proves that $\mathfrak{I}_2 \subset \mathfrak{I}_1$.

Hence $\mathfrak{I}_1 = \mathfrak{I}_2$ and the proof is completed.

11.42. Remark. We conclude from 11.41 that for a set of continuous mappings, the compact-index topology depends only on the topology of the range space.

11.43. Theorem. *Let X be a locally compact space, let Y be a topological space, let Φ be a set of continuous mappings of X into Y and let $\pi : X \times \Phi \to Y$ be defined by $(x, \varphi)\pi = x\varphi$ $(x \in X, \varphi \in \Phi)$. Then the compact-open topology of Φ is the least topology of Φ which makes π continuous.*

Proof. Let $\mathfrak{I}$ be the compact-open topology of Φ.

We show that $\mathfrak{I}$ makes π continuous. Let $x \in X$, let $\varphi \in \Phi$ and let W be an open neighborhood of $(x, \varphi) = x\varphi$. There exists a compact neighborhood U of x such that $U\varphi \subset W$. Define $V = (U, W)_\Phi$. Now V is a $\mathfrak{I}$-neighborhood of φ and $(U \times V)\pi \subset W$. This shows that $\mathfrak{I}$ makes π continuous.

Now let $\mathfrak{I}_0$ be a topology of Φ which makes π continuous. We show $\mathfrak{I} \subset \mathfrak{I}_0$. Let A be a compact subset of X, let E be an open subset of Y and let $\varphi \in (A, E)_\Phi$. It is enough to show that $U \subset (A, E)_\Phi$ for some $\mathfrak{I}_0$-neighborhood U of φ. Since $(A \times [\varphi])\pi \subset E$ and A is compact, there exists a $\mathfrak{I}_0$-neighborhood U of φ such that $(A \times U)\pi \subset E$ whence $U \subset (A, E)_\Phi$. The proof is completed.

11.44. Definition. Let X be a topological space, let $\mathfrak{A}$ be the class of all compact subsets of X, let $\mathcal{E}$ be the class of all open subsets of X and let Φ be a set of permutations of X. For $A, E \subset X$ define

$$(A, E)_\Phi = [\varphi \mid \varphi \in \Phi \text{ and } A\varphi \subset E],$$

$$(A, E)^*_\Phi = [\varphi \mid \varphi \in \Phi \text{ and } A\varphi^{-1} \subset E].$$

Define

$$\mathcal{S} = [(A, E)_\Phi \mid A \in \mathcal{A} \text{ and } E \in \mathcal{E}],$$

$$\mathcal{S}^* = [(A, E)^*_\Phi \mid A \in \mathcal{A} \text{ and } E \in \mathcal{E}],$$

$$\hat{\mathcal{S}} = \mathcal{S} \cup \mathcal{S}^*.$$

The topologies $\mathfrak{I}$, $\mathfrak{I}^*$, $\hat{\mathfrak{I}}$ of Φ generated by $\mathcal{S}$, $\mathcal{S}^*$, $\hat{\mathcal{S}}$ are called the *compact-open*, the *inverse compact-open*, the *bilateral compact-open* topologies of Φ. It is clear that:

(1) If Φ is a group, then $\mathfrak{I}^*$ is the image of $\mathfrak{I}$ under the group inversion of Φ.
(2) $\hat{\mathfrak{I}} = \mathfrak{I} \vee \mathfrak{I}^*$.

11.45. THEOREM. *Let X be a uniform space and let Φ be a group of homeomorphisms of X onto X. Then the* {*compact-index*} {*inverse compact-index*} {*bilateral compact-index*} *topology of Φ coincides with the* {*compact-open*} {*inverse compact-open*} {*bilateral compact-open*} *topology of Φ.*

PROOF. Use 11.41.

11.46. THEOREM. *Let X be a locally compact T_2-space and let Φ be a group of homeomorphisms of X. Then the bilateral compact-open topology of Φ is the least topology of Φ which makes Φ a topological homeomorphism group of X.*

PROOF. Use 11.43 and 11.36.

11.47. NOTES AND REFERENCES.

The purpose of this section is to set forth those developments of the theory of function spaces needed elsewhere in the book. It can be read independently of the other sections. Most of the results can be found in Bourbaki [4], where some references to the pertinent literature can be found. See also Arens [1, 2].

PART II. THE MODELS

12. SYMBOLIC DYNAMICS

12.01. STANDING NOTATION. Let S be a finite set which contains more than one element and let S be called the *symbol class*. We shall use $i, j, k, m, n, p, q, r, s, t$ as integer variables, that is variables ranging over $\mathcal{I}$.

12.02. DEFINITION. A $\{$*right*$\}\{$*left*$\}$ *ray* is a subset R of $\mathcal{I}$ such that $\{R = [i \mid p \leqq i]\}\{R = [i \mid i \leqq p]\}$ for some $p \in \mathcal{I}$. An *interval* is a subset I of $\mathcal{I}$ such that $I = [i \mid p \leqq i \leqq q]$ for some $p, q \in \mathcal{I}$ with $p \leqq q$. If $p \in \mathcal{I}$, then $\{p, +\infty\}$, $\{-\infty, p\}$ denote the rays $[i \mid p \leqq i]$, $[i \mid i \leqq p]$. If $p, q \in \mathcal{I}$ with $p \leqq q$, then $\{p, q\}$ denotes the interval $[i \mid p \leqq i \leqq q]$. We also write $\{-\infty, +\infty\} = \mathcal{I}$.

12.03. DEFINITION. A $\{$*bisequence*$\}\{$*right sequence*$\}\{$*left sequence*$\}\{$*block*$\}$ is a function on $\{\mathcal{I}\}\{$a right ray$\}\{$a left ray$\}\{$an interval$\}$ to S.

12.04. DEFINITION. We make the following definitions:

(1) If A is a {bisequence}{right sequence}{left sequence}, then the *reverse of* A, denoted $A^{\smile}$ or $\breve{A}$, is the {bisequence}{left sequence}{right sequence} B such that dmn $B = -$dmn A and $B(-i) = A(i)$ ($i \in$ dmn A).

(2) If A is a block, then the *reverse of* A, denoted $A^{\smile}$ or $\breve{A}$, is the block B such that dmn $B =$ dmn A and $B(p + q - i) = A(i)$ ($i \in$ dmn A) where p, q are the first, last elements of dmn A.

(3) If A is a {bisequence}{right sequence}{left sequence}{block} and if $n \in \mathcal{I}$, then the *n-translate of* A, denoted A^n, is the {bisequence}{right sequence}{left sequence}{block} B such that dmn $B = n +$ dmn A and $B(n + i) = A(i)$ ($i \in$ dmn A).

(4) If A and B are {bisequences}{right sequences}{left sequences}{blocks}, then A is *similar to* B, this statement being denoted $A \sim B$, in case there exists $n \in \mathcal{I}$ such that $A^n = B$.

(5) If A is a right sequence or left sequence or block and if B is a bisequence or right sequence or left sequence or block, then A is *contained in* B and B *contains* A in case A is a restriction of B or equivalently B is an extension of A; in the event that A is a {right sequence}{left sequence}{block} we may also say that A is a $\{$*right subsequence*$\}\{$*left subsequence*$\}\{$*subblock*$\}$ *of* B.

(6) If A is a {right sequence}{left sequence}{block} and if B is a bisequence or right sequence or left sequence or block, then A *appears in* B in case A is similar to a {right subsequence}{left subsequence}{subblock} of B.

(7) If A is a block, then the *length of* A is the cardinal of dmn A.

(8) If $n > 0$, then an *n-block* is a block of length n.

12.05. DEFINITION. Let $A_0, A_1, \cdots$ be a sequence of blocks such that each A_n ($n \geqq 0$) is contained in A_{n+1}. The *union of* $A_0, A_1, \cdots$, denoted

$\bigcup_{n=0}^{+\infty} A_n$, is the bisequence or right sequence or left sequence or block A such that dmn $A = \bigcup_{n=0}^{+\infty}$ dmn A_n and each $A_n (n \geqq 0)$ is contained in A.

Variations of this notation are obvious, for example, the union of a finite class of blocks.

12.06. DEFINITION. Let A be a {right sequence}{block}. Then $\dot{A}$ denotes A^{-n} where n is the least element of dmn A.

Let A, B be blocks with lengths n, m. Then $\dot{A}B$ denotes the $(n + m)$-block C such that the initial n-subblock of C is $\dot{A}$ and the terminal m-subblock of C is similar to B. Analogously $A\dot{B}$ denotes the $(n + m)$-block D such that the terminal m-subblock of D is $\dot{B}$ and the initial n-subblock of D is similar to A.

The meaning of $A_1 A_2 \cdots \dot{A}_p \cdots A_{n-1} A_n$ is now clear where $A_1, \cdots, A_n$ are blocks except that A_1 may be a left sequence (if $p > 1$) and A_n may be a right sequence.

Other uses of this "indexing-by-dot" notation are obvious. For example, if $\cdots, A_{-1}, A_0, A_1, \cdots$ are blocks, then $(\cdots A_{-1}\dot{A}_0 A_1 A_2 A_3 \cdots)$ denotes a certain bisequence. If x is a bisequence, then we may write $x = (\cdots x(-1)\dot{x}(0)x(1) \cdots)$. If, a, b, $c \in S$, then $\dot{a}$ denotes a certain 1-block, $a\dot{b}$ denotes a certain 2-block, $ab\dot{c}$ denotes a certain 3-block, etc.

12.07. DEFINITION. Let A, B be blocks with lengths n, m. Then $\hat{A}B$ denotes the $(n + m)$-block C such that the initial n-subblock of C is A and the terminal m-subblock of C is similar to B. Analogously $A\hat{B}$ denotes the $(n + m)$-block D such that the terminal m-subblock of D is B and the initial n-subblock of D is similar to A.

Other uses of this "indexing-by-roof" notation are obvious. For example, if $A_1, \cdots, A_n$ are blocks, then $A_1 A_2 \cdots \hat{A}_p \cdots A_{n-1} A_n$ denotes a certain block which contains A_p.

12.08. DEFINITION. The *bisequence space* is the set of all bisequences, that is, the set $S^{\mathscr{I}}$. We consider the symbol class S to be provided with its discrete topology and we consider the bisequence space $S^{\mathscr{I}}$ to be provided with its product topology or equivalently its point-open topology.

12.09. STANDING NOTATION. Let X denote the bisequence space $S^{\mathscr{I}}$. Let $\rho : X \times X \rightarrow \mathcal{R}$ be the function defined by $\rho(x, y) = (1 + \sup [n \mid x(i) = y(i)$ for $\mid i \mid < n])^{-1}$ $(x, y \in X)$.

12.10. REMARK. The bisequence space X is a self-dense zero-dimensional compact metrizable space with ρ as compatible metric and indeed X is homeomorphic to the Cantor discontinuum.

12.11. DEFINITION. The *shift transformation of* X is the homeomorphism

$$\sigma : X \overset{\text{onto}}{\rightarrow} X$$

defined by $x\sigma = (x(i + 1) \mid i \in I)$ $(x \in X)$. The *symbolic flow generated by* S is the discrete flow on the bisequence space X generated by the shift transformation σ of X and is denoted by (S, X, σ). We shall call S the *symbol class* of (S, X, σ).

12.12. REMARK. Let $x \in X$ and let σ be the shift transformation of X. Then

$$x\sigma = (\cdots\ x(-1)\dot{x}(0)x(1)\ \cdots)\sigma = (\cdots\ x(0)\dot{x}(1)x(2)\ \cdots).$$

12.13. REMARK. The symbolic flow generated by S {coincides}{is isomorphic} with the {left}{right} functional transformation group over $\mathscr{I}$ to S.

12.14. REMARK. Let T be a set such that crd S = crd T. Then the symbolic flow generated by S is isomorphic to the symbolic flow generated by T.

12.15. DEFINITION. Let n = crd S. By virtue of 12.14 the symbolic flow generated by S may be called the *symbolic flow on n symbols.*

12.16. STANDING NOTATION. Let (S, X, σ) denote the symbolic flow generated by S. Properties of invariance, recursion, etc., are relative to the shift transformation.

12.17. REMARK. Let $x \in X$. Then:

(1) x is periodic if and only if there exists $p > 0$ such that $x(i + p) = x(i)$ $(i \in \mathscr{I})$.

(2) If x is periodic, then the period of x is the least $p > 0$ such that $x(i + p) = x(i)$ $(i \in \mathscr{I})$.

12.18. REMARK. Let $x \in X$. Then the following statements are pairwise equivalent:

(1) x is regularly almost periodic.

(2) If $n > 0$, then there exists $p > 0$ such that $x(i + pj) = x(i)$ $(|\,i\,| \leqq n, j \in \mathscr{I})$.

(3) If $n \in \mathscr{I}$, then there exists $p > 0$ such that $x(n + pj) = x(n)$ $(j \in \mathscr{I})$.

12.19. REMARK. Let $x \in X$. Then the following statements are equivalent:

(1) x is isochronous.

(2) If $n > 0$, then there exist $p > 0$ and $q \in \mathscr{I}$ such that $x(i + pj + q) = x(i)$ $(|\,i\,| \leqq n, j \in \mathscr{I})$.

12.20. REMARK. Let $x \in X$. Then the following statements are pairwise equivalent:

(1) x is almost periodic.

(2) If $n > 0$, then there exists a syndetic subset E of $\mathscr{I}$ such that $x(i + j) = x(i)$ $(|\,i\,| \leqq n, j \in E)$.

(3) If A is a subblock of x, then there exists $k > 0$ such that A appears in every k-subblock of x.

(4) If $n > 0$, then there exists $k > 0$ such that every n-subblock of x appears in every k-subblock of x.

12.21. REMARK. Let $x \in X$. Then the following statements are equivalent:

(1) x is recurrent.

(2) If $n > 0$, then there exists an extensive subset E of $\mathcal{I}$ such that $x(i + j) = x(i)$ ($|i| \leqq n, j \in E$).

12.22. Remark. Let $x \in X$. Then the following statements are equivalent:

(1) x is transitive.

(2) Every block appears in x.

12.23. Remark. Let $x \in X$. Then the following statements are equivalent:

(1) x is extensively transitive.

(2) If $x = B\dot{A}$, then every block appears in both A and B.

12.24. Remark. The following statements are valid:

(1) The set of all periodic points is an invariant countable dense subset of X.

(2) σ is regionally regularly almost periodic, regionally almost periodic and regionally recurrent.

(3) The set of all recurrent points is an invariant residual G_δ subset of X.

(4) The set of all extensively transitive points is an invariant residual G_δ subset of X.

(5) σ is mixing.

(6) σ is expansive.

12.25. Remark. Let $x, y \in X$. Then the following statements are equivalent:

(1) x and y are {positively}{negatively}{doubly} asymptotic.

(2) There exists $n > 0$ such that $\{x(i) = y(i)\ (i \geqq n)\}\ \{x(i) = y(i)\ (i \leqq -n)\}\ \{x(i) = y(i)\ (|i| \geqq n)\}$.

12.26. Standing notation. Let $S = [0, 1]$. Thus (S, X, σ) denotes the symbolic flow on 2 symbols.

12.27. Definition. The *dual of* $\{0\}\{1\}$, denoted $\{0'\}\{1'\}$, is $\{1\}\{0\}$.

Let A be a {bisequence}{right sequence}{left sequence}{block}. The *dual of* A, denoted A', is the {bisequence}{right sequence}{left sequence}{block} such that dmn A' = dmn A and $A'(i) = A(i)'$ ($i \in$ dmn A).

Let δ denote the homeomorphism of X onto X defined by $x\delta = x'$ ($x \in X$).

12.28. Definition. Define the sequence $Q_0, Q_1, \cdots$ of blocks inductively as follows:

(1) $Q_0 = \dot{0}$.

(2) If $n \geqq 0$, then $Q_{n+1} = \dot{Q}_n Q'_n$.

Define $Q = \bigcup_{n=0}^{+\infty} Q_n$. Define $\mu = \breve{Q}\dot{Q} \in X$.

12.29. Definition. Let A be a {bisequence}{right sequence}{left sequence}{block} and let dmn $A = \{p, q\}$. The *Q_2-extension* of A, denoted A^*, is the {bisequence}{right sequence}{left sequence}{block} such that dmn $A^* = \{2p, 2q + 1\}$, $A^*(2i) = A(i)$ ($p \leqq i \leqq q$) and $A^*(2i + 1) = A'(i)$ ($p \leqq i \leqq q$).

12.30. Remark. Let $n \geqq 0$. Then:

(1) Q_n is the initial 2^n-subblock of Q_{n+p} ($p \geqq 0$).

(2) $\breve{Q}_n = Q_n$ if n is even.
$\breve{Q}_n = Q'_n$ if n is odd.

(3) $Q^*_n = Q_{n+1}$.

(4) $Q^* = Q$.

(5) $Q_{n+3} = \dot{Q}_nQ'_nQ'_nQ_nQ'_nQ_nQ_nQ'_n$
$Q'_{n+3} = \dot{Q}'_nQ_nQ_nQ'_nQ_nQ'_nQ'_nQ_n$.

(6) $\breve{\mu} = \mu$.

(7) $\mu \mid \{-2^n, -1\} \sim Q_n$ if n is even.
$\mu \mid \{-2^n, -1\} \sim Q'_n$ if n is odd.

(8) $\mu \mid \{-2^n, 2^n - 1\} = Q_n\dot{Q}_n$ if n is even.
$\mu \mid \{-2^n, 2^n - 1\} = Q'_n\dot{Q}_n$ if n is odd.

(9) If n is even and if P_i $(i \in \mathcal{I})$ are blocks such that

$$P_i = \begin{cases} Q_n & \text{if } \mu(i) = 0 \\ Q'_n & \text{if } \mu(i) = 1, \end{cases}$$

then $\mu = (\cdots P_{-1}\dot{P}_0P_1 \cdots)$.

(10) If n is odd and if P_i $(i \in \mathcal{I})$ are blocks such that

$$P_i = \begin{cases} Q_n & \text{if } \mu(i) = 0 \text{ and } i \geqq 0, \\ Q'_n & \text{if } \mu(i) = 0 \text{ and } i < 0, \\ Q'_n & \text{if } \mu(i) = 1 \text{ and } i \geqq 0, \\ Q_n & \text{if } \mu(i) = 1 \text{ and } i < 0, \end{cases}$$

then $\mu = (\cdots P_{-1}\dot{P}_0P_1 \cdots)$.

12.31. Remark. For $n > 0$ define $\theta(n) = \sum_{i=0}^{k} a_i$ where $n = \sum_{i=0}^{k} a_i 2^i$, $a_i = 0$ or 1 $(i = 0, \cdots, k-1)$ and $a_k = 1$, and define $\theta(0) = 0$. If $n \geqq 0$, then $\mu(n) = \theta(n) \pmod 2$.

Proof. The statement is true for $n = 0$ or 1, that is, for n such that $0 \leqq n \leqq 2^p - 1$ where $p = 1$. Let $p \geqq 1$ and assume the statement is true for n such that $0 \leqq n \leqq 2^p - 1$. We show the statement is true for n such that $0 \leqq n \leqq 2^{p+1} - 1$. Let n be such that $2^p - 1 < n \leqq 2^{p+1} - 1$. It is enough to show that $\mu(n) = \theta(n) \pmod 2$. Since $\dot{\mu}(0) \cdots \mu(2^{p+1} - 1) = Q_{p+1} = \dot{Q}_pQ'_p$, we have $\mu(n) = \mu'(n - 2^p) \neq_2 \mu(n - 2^p) \equiv_2 \theta(n - 2^p) \neq_2 \theta(n)$. Hence $\mu(n) = \theta(n) \pmod 2$. The proof is completed.

12.32. Theorem. *μ is almost periodic.*

Proof. Let A be a subblock of μ. Choose a positive even integer n such that A appears in $Q_n\dot{Q}_n$. Let $k = 2^{n+5}$ and let B be a k-subblock of μ. We show that A appears in B. By 12.30 (9), B contains a subblock similar to Q_{n+3} or $\dot{Q}'_{n+3}$. By 12.30 (5), we have that B contains a subblock similar to $Q_n\dot{Q}_n$ and hence a subblock similar to A. The proof is completed.

12.33. DEFINITION. Define M to be the orbit-closure of μ under σ.

12.34. REMARK. The following statements are valid:
(1) M is a minimal orbit-closure under σ.
(2) If A is a subblock of μ, then $\breve{A}$, A', A^* appear in μ.
(3) If $x \in M$, then $\breve{x}$, x', $x^* \in M$.

12.35. REMARK. The three unary operations $\breve{}$, $'$, $*$ applied successively to μ yield exactly four different bisequences, namely, μ, μ', μ^*, $\mu^{*\prime}$.

12.36. DEFINITION. Define $\nu = \mu^*$ whence $\mu = \nu^*$.

12.37. REMARK. In summary:

$$\mu = \breve{Q}\dot{Q},$$

$$\mu' = \breve{Q}'\dot{Q}',$$

$$\nu = \breve{Q}'\dot{Q},$$

$$\nu' = \breve{Q}\dot{Q}'$$

are the only bisequences obtained from μ by use of $\breve{}$, $'$, $*$.

12.38. REMARK. The following statements are valid:
(1) $\mu, \mu', \nu, \nu' \in M$.
(2) μ and μ' are separated.
ν and ν' are separated.
μ and ν are negatively separated and positively asymptotic.
μ' and ν' are negatively separated and positively asymptotic.
μ and ν' are negatively asymptotic and positively separated.
μ' and ν are negatively asymptotic and positively separated.

12.39. THEOREM. *$\sigma \mid M$ is not {locally recurrent} {locally almost periodic} {isochronous} at any point of M.*

12.40. REMARK. Let $n = 2^k > 0$ and let A be the class of all orbit-closures under $\sigma^n \mid M$. Then crd $A = n$. Hence M is not totally minimal under $\sigma \mid M$ and $\sigma \mid M$ is not mixing.

12.41. REMARK. The set of all nonisochronous almost periodic points of X is dense in X.

PROOF. Let A be a block. Construct $(\cdots P_{-1}\hat{P}_0P_1 \cdots)$ where

$$P_i = \begin{cases} A & \text{if } \mu(i) = 0, \\ A' & \text{if } \mu(i) = 1, \end{cases} \qquad (i \in \mathcal{I}).$$

12.42. DEFINITION. Let x be a bisequence and let A be a block. An *A-representation of x* is a family $(A_i \mid i \in \mathcal{I})$ of blocks such that $A_i = A$ or $A'(i \neq 0)$, $A_0 \sim A$ or A', $0 \in$ dmn A_0 and $x = (\cdots A_{-1}\hat{A}_0A_1 \cdots)$.

12.43. DEFINITION. Let $n > 0$ and let x be a bisequence which has a Q_n-representation $(A_i \mid i \in \mathcal{I})$. The Q_{n-1}-representation of x *induced by* $(A_i \mid i \in \mathcal{I})$ is the unique Q_{n-1}-representation $(B_i \mid i \in \mathcal{I})$ of x such that $A_0 = \hat{B}_0B_1$ or $B_{-1}\hat{B}_0$.

12.44. LEMMA. *Let $x \in M$ and let $n \geqq 0$. Then there exists exactly one Q_n-representation of x.*

PROOF. By 12.30(9 & 10) there exists a Q_n-representation of μ and therefore of $\mu\sigma^i$ $(i \in \mathcal{I})$. Since $\bigcup_{i \in \mathcal{I}} \mu\sigma^i$ is dense in M, it follows that x has at least one Q_n-representation.

For $n \geqq 0$ let S_n denote the statement that there exists at most one Q_n-representation of x. Clearly S_0 is true. Let $k \geqq 0$. Assume S_k is true. We show that S_{k+1} is true. Let $(A_i \mid i \in \mathcal{I})$ and $(B_i \mid i \in \mathcal{I})$ be Q_{k+1}-representations of x. In order to show that $(A_i \mid i \in \mathcal{I}) = (B_i \mid i \in \mathcal{I})$ it is enough to show that $p = r$ where $\{p, q\} =$ dmn A_0 and $\{r, s\} =$ dmn B_0. We may suppose that $p \leqq r \leqq q \leqq s$. Assume that $p \neq r$ whence $p < r$. We seek a contradiction. If $r \neq p + 2^k$, then the Q_k-representations of x induced by $(A_i \mid i \in \mathcal{I})$ and $(B_i \mid i \in \mathcal{I})$ are not identical which is contrary to S_k. Hence $r = p + 2^k$. Each A_i, B_i $(i \in \mathcal{I})$ is similar to $Q_{k+1} = \dot{Q}_kQ'_k$ or to $Q'_{k+1} = Q_kQ_k$. Consideration of the common Q_k-representation of x induced by $(A_i \mid i \in \mathcal{I})$ and $(B_i \mid i \in \mathcal{I})$ now shows that x is periodic. This is a contradiction. The proof is completed.

12.45. LEMMA. *Let $x \in M$ and let A, B be left, right sequences such that $x = \hat{A}B$ or $A\hat{B}$. Then:*

(1) *If $A \sim \breve{Q}$ or $\breve{Q}'$, then $B \sim Q$ or Q'.*

(2) *If $B \sim Q$ or Q', then $A \sim \breve{Q}$ or $\breve{Q}'$.*

PROOF. Let $x \in M$ and let P be a left sequence such that $x = P\dot{Q}$. By 12.34, it is enough to show that $P = \breve{Q}$ or $\breve{Q}'$. Let $n \geqq 0$. It is enough to show that $x \mid \{-2^n, -1\} \sim Q_n$ or Q'_n since the end elements of $\{Q_n\}\{Q'_n\}$ are $\{0\text{'s}\}\{1\text{'s}\}$ when n is even. To show this it is enough to prove that if $(A_i \mid i \in \mathcal{I})$ is the Q_n-representation of x, then 0 is the least element of dmn A_0. Let $(A_i \mid i \in \mathcal{I})$ be the Q_n-representation of x and let $(B_i \mid i \in \mathcal{I})$ be the Q_n-representation of $\mu = \breve{Q}\dot{Q}$. Now 0 is the least element of dmn B_0. Choose $y \in M$ such that for some sequence $n_1, n_2, \cdots$ of positive integers we have

$$\lim_{i \to \infty} x\sigma^{n_i} = \lim_{i \to \infty} \mu\sigma^{n_i} = y.$$

It follows that if 0 is not the least element of dmn A_0, then y has two different Q_n-representations. The proof is completed.

12.46. DEFINITION. Let $x \in M$, let $n > 0$ and let $(A_i \mid i \in \mathcal{I})$ be the Q_n-representation of x. Then $(A_i \mid i \in \mathcal{I})$ is $\{$*left*$\}\{$*right*$\}$ *indexed* provided that $\{A_0 = \hat{B}_0B_1\}\{A_0 = B_{-1}\hat{B}_0\}$ where $(B_i \mid i \in I)$ is the Q_{n-1}-representation of x.

12.47. DEFINITION. Let $a = (a_0, a_1, \cdots)$ be a dyadic sequence, that is, $a_n = 0$ or 1 $(n = 0, 1, \cdots)$. Define the sequence $F_0, F_1, \cdots$ of blocks inductively as follows:

(1) $F_0 = \dot{a}_0$.

(2) If $n \geqq 0$, then

$$F_{n+1} = \begin{cases} \hat{F}_n F'_n & \text{if } a_{n+1} = 0, \\ F'_n \hat{F}_n & \text{if } a_{n+1} = 1. \end{cases}$$

Define $F_a = \bigcup_{n=0}^{+\infty} F_n$.

12.48. REMARK. Let a be a dyadic sequence. Then:

(1) $F_n \sim Q_n$ or Q'_n $(n \geqq 0)$ in the notation of 12.28.

(2) Every subblock of $\{F_a\}\{\mu\}$ appears in $\{\mu\}\{F_a\}$.

(3) If a contains infinitely many 0's and infinitely many 1's, then F_a is a bisequence such that $F_a \in M$.

(4) If a contains only finitely many {0's}{1's}, then $\{F_a \sim \check{Q}$ or $\check{Q}'\}\{F_a \sim Q$ or $Q'\}$.

(5) If $a = (0, a_1, \cdots)$ and if $b = (1, a_1, \cdots)$, then $F_a = F'_b$.

12.49. THEOREM. *Let N be the set of all bisequences x such that $x = F_a$ or $\hat{F}_a \check{F}_a$ or $\hat{F}_a \check{F}'_a$ or $\check{F}_a \hat{F}_a$ or $\check{F}'_a \hat{F}_a$ for some dyadic sequence a. Then $N = M$.*

PROOF. That $N \subset M$ follows immediately from 12.48.

We show $M \subset N$. Let $x \in M$. Define the dyadic sequence $a = (a_0, a_1, \cdots)$ as follows:

$a_0 = x(0)$.

$$a_i = \begin{Bmatrix} 0 \\ 1 \end{Bmatrix} \text{ if the } Q_i\text{-representation of } x \text{ is } \begin{Bmatrix} \text{left} \\ \text{right} \end{Bmatrix} \text{ indexed } (i > 0).$$

We adopt the notation $F_0, F_1, \cdots$ of 12.47. For $n \geqq 0$ let $(A_i^n \mid i \in \mathcal{I})$ be the Q_n-representation of x.

We first show that $F_n = A_0^n$ $(n \geqq 0)$. Clearly $F_0 = A_0^0$. Let $k \geqq 0$. Assume $F_k = A_0^k$. We show $F_{k+1} = A_0^{k+1}$. If $a_{k+1} = 0$, then $A_0^{k+1} = \hat{A}_0^k A_1^k = \hat{A}_0^k A_0^{k\prime} = \hat{F}_k F'_k = F_{k+1}$. If $a_{k+1} = 1$, then $A_0^{k+1} = A_{-1}^k \hat{A}_0^k = A_0^{k\prime} \hat{A}_0^k = F'_k \hat{F}_k = F_{k+1}$. This completes the proof that $F_n = A_0^n$ $(n \geqq 0)$.

We now have that F_a is contained in x since $F_a = \bigcup_{n=0}^{+\infty} F_n = \bigcup_{n=0}^{+\infty} A_0^n$ and $\bigcup_{n=0}^{+\infty} A_0^n$ is contained in x.

Case I. The sequence a contains infinitely many 0's and 1's. Since F_a is a bisequence, $x = F_a$ and $x \in N$.

Case II. The sequence a contains only finitely many 0's. Then F_a is a left sequence such that $F_a \sim \check{Q}$ or $\check{Q}'$. Now $x = \hat{F}_a P$ for some right sequence P. By 12.45, $P \sim Q$ or Q'. Therefore $P \sim \check{F}_a$ or $\check{F}'_a$, $x = \hat{F}_a \check{F}_a$ or $\hat{F}_a \check{F}'_a$, and $x \in N$.

Case III. The sequence a contains only finitely many 1's. Then F_a is a right sequence such that $F_a \sim Q$ or Q'. Now $x = P\hat{F}_a$ for some left sequence P. By

12.45, $P \sim \check{Q}$ or $\check{Q}'$. Therefore $P \sim \check{F}_a$ or $\check{F}'_a$, $x = \check{F}_a\hat{F}_a$ or $\check{F}'_a\hat{F}_a$, and $x \in N$. The proof is completed.

12.50. LEMMA. *Let $x \in M$ such that x does not belong to the union of the orbits of μ, μ', ν, ν' and let $n > 0$. Then there exists $m > 0$ such that if $(A_i \mid i \in \mathfrak{s})$ is the Q_m-representation of x, then* $\{-n, n\} \subset \operatorname{dmn} A_0$.

PROOF. Cf. the proof of 12.49.

12.51. DEFINITION. Define $\pi : X \to X$ by

$$(x\pi)(i) = \begin{cases} 0 & \text{if } x(i-1) \neq x(i), \\ 1 & \text{if } x(i-1) = x(i), \end{cases} \qquad (x \in X, i \in \mathfrak{s}).$$

Define

$$\eta = \mu\pi,$$

$$\zeta = \nu\pi,$$

$$H = M\pi.$$

12.52. REMARK. It follows from 12.31 that $\eta(0) = 1$, $\zeta(0) = 0$, and

$$\eta(i) = \zeta(i) = \begin{cases} 0 & \text{if } k \text{ is even}, \\ 1 & \text{if } k \text{ is odd}, \end{cases} \qquad (i, j, k \in \mathfrak{s}; i \neq 0; i = j2^k; j \text{ odd}).$$

12.53. NOTATION. Let $x, y \in X$. The statement that x and y are {negatively} {positively} {doubly} asymptotic is denoted by $\{x \downarrow y\}\{x \uparrow y\}\{x \updownarrow y\}$.

12.54. REMARK. Let $x, y \in X$. Then:

(1) π is an exactly 2-to-1 continuous-open mapping of X onto X.

(2) $x\pi = x'\pi$.
$x\pi\pi^{-1} = [x, x'] = [x, x\delta]$.

(3) $x\sigma\pi = x\pi\sigma$.

(4) If x is recursive, then $x\pi$ is recursive.

(5) If $x \downarrow y$, then $x\pi \downarrow y\pi$.
If $x \uparrow y$, then $x\pi \uparrow y\pi$.

(6) $\mu\pi = \eta = \mu'\pi$, $\eta\pi^{-1} = [\mu, \mu']$.
$\nu\pi = \zeta = \nu'\pi$, $\zeta\pi^{-1} = [\nu, \nu']$.

(7) $\eta(0) = 1$.
$\zeta(0) = 0$.
$\eta(i) = \zeta(i)$ $(i \neq 0)$.

(8) $\eta \updownarrow \zeta$ and hence neither η nor ζ is regularly almost periodic.

(9) $\eta, \zeta \in H$.

(10) H is a minimal orbit-closure under σ.

12.55. THEOREM. *The following statements are valid:*

(1) *η and ζ are isochronous.*

(2) *The set of all regularly almost periodic points of* H *coincides with the complement in* H *of the union of the orbits of* η *and* ζ.

(3) *The orbits of* η *and* ζ *are the only nonseparated orbits in* H.

(4) *If* φ *is a homeomorphism of* H *onto* H *such that* $\sigma\varphi = \varphi\sigma$, *then* $\varphi = \sigma^n$ *for some integer* n.

(5) *The only nondegenerate traces under* $\sigma \mid H$ *are* $[\eta\sigma^n, \zeta\sigma^n]$ $(n \in \mathscr{I})$.

PROOF. Let $x \in M$ such that x does not belong to the union of the orbits of μ, μ', ν, ν'. It is enough to show that $y = x\pi$ is regularly almost periodic. Let $n > 0$. By 12.50 there exists $m > 0$ such that if $(A_i \mid i \in \mathscr{I})$ is the Q_m-representation of x and if dmn $A_0 = \{p, q\}$, then $\{-n, n\} \subset \{p + 1, q\}$. Let $(A_i \mid i \in \mathscr{I}$ be the Q_m-representation of x and let dmn $A_0 = \{p, q\}$. If $j \in \mathscr{I}$, then either $x(i + 2^m j) = x(i)$ $(i \in \{p, q\})$ or $x(i + 2^m j) = x(i)'$ $(i \in \{p, q\})$. Hence $y(i + 2^m j) = y(i)$ $(i \in \{p + 1, q\}, j \in \mathscr{I}$ and $y(i + 2^m j) = y(i)$ $(|\, i\,| \leqq n$, $j \in \mathscr{I})$. The proof is completed.

12.56. THEOREM. *The following statements are valid:*

(1) *The only nonseparated orbits in* M *are the obvious ones, namely, the orbits of* μ *and* ν, μ' *and* ν', μ *and* ν', μ' *and* ν.

(2) *If* φ *is a homeomorphism of* M *onto* M *such that* $\sigma\varphi = \varphi\sigma$, *then* $\varphi = \sigma^n\delta^m$ *for some integers* n, m $(m = 0, 1)$.

(3) *Let* E *be the union of the orbits of* μ, μ', ν, ν'. *Then the traces under* $\sigma \mid M$ *are:*

$$[x, x'] \qquad (x \in M - E),$$

$$[\mu\sigma^n, \mu'\sigma^n, \nu\sigma^n, \nu'\sigma^n] \qquad (n \in \mathscr{I}).$$

12.57. DEFINITION. Let C be an oriented circle (1-sphere) of circumference 1 and let C be provided with its natural topology. Let α be an irrational number such that $0 < \alpha < 1$. Let A be a closed interval in C of length α and let $\{a^-\}\{a^+\}$ be the $\{$first$\}\{$last$\}$ endpoint of A. Let $A^- = A - a^+$ and let $A^+ = A - a^-$. Let $\rho : C \to C$ be the rotation of C such that $a^-\rho = a^+$. For $p \in C$ define x_p^-, $x_p^+ \in X$ as follows:

$$x_p^-(i) = \begin{cases} 0 & \text{if } p\rho^i \in A^-, \\ 1 & \text{if } p\rho^i \notin A^-, \end{cases} \qquad (i \in \mathscr{I})$$

$$x_p^+(i) = \begin{cases} 0 & \text{if } p\rho^i \in A^+, \\ 1 & \text{if } p\rho^i \notin A^+, \end{cases} \qquad (i \in \mathscr{I}).$$

12.58. REMARK. Let $p, q \in C$. Then:

(1) The following statements are pairwise equivalent:

(I) $x_p^- = x_p^+$.

(II) $p\rho^n \neq a^-$ $\quad (n \in \mathscr{I})$.

(III) $p\rho^n \neq a^+$ $\quad (n \in \mathscr{I})$.

(2) If $n \in \mathcal{I}$ and if $p\rho^n = a^-$, then:

(I) $x_p^-(i) = x_p^+(i) \qquad (i \in \mathcal{I}; i \neq n, n+1)$.

(II) $x_p^-(n) = 0, \qquad x_p^-(n+1) = 1,$

$x_p^+(n) = 1, \qquad x_p^+(n+1) = 0.$

(III) $x_p^- \updownarrow x_p^+$.

(3) If $n \in \mathcal{I}$ and if $p\rho^n = a^+$, then:

(I) $x_p^-(i) = x_p^+(i) \qquad (i \in \mathcal{I}; i \neq n-1, n)$.

(II) $x_p^-(n-1) = 0, \qquad x_p^-(n) = 1,$

$x_p^+(n-1) = 1, \qquad x_p^+(n+1) = 0.$

(III) $x_p^- \updownarrow x_p^+$.

(4) $[x_p^-, x_p^+] \cap [x_q^-, x_q^+] = \emptyset$ if and only if $p \neq q$.

12.59. Definition. For $D \subset C$ define $D^* = [x_p^- \mid p \in D] \cup [x_p^+ \mid p \in D]$. Define $\Gamma = C^*$. Define $\mathfrak{J}$ to be the class of all subsets E of Γ such that:

(1) If $p \in C$ such that $x_p^- = x_p^+ \in E$, then $D^* \subset E$ for some open interval D in C which contains p.

(2–3) If $p \in C$ such that $x_p^- \neq x_p^+$ and $\{x_p^- \in E\}\{x_p^+ \in E\}$, then $D^* \subset E$ for some nondegenerate closed interval D in C with {initial}{terminal} endpoint p.

12.60. Notation. Let $p, p_1, p_2, \cdots \in C$. Then $\{(-) \lim_{i\to\infty} p_i = p\}$ $\{(+) \lim_{i\to\infty} p_i = p\}$ means that if D is a nondegenerate closed interval in C with {initial}{terminal} endpoint p, then $p_i \in D$ for all sufficiently large i. The symbol x_p^* denotes x_p^- or x_p^+ .

12.61. Remark. Let $p, p_1, p_2, \cdots \in C$. Then:

(1) If $\lim_{i\to+\infty} p_i = p$ and if $x_p^- = x_p^+$, then

$$\lim_{i\to+\infty} x_{p_i}^- = x_p^- \text{ in } X,$$

$$\lim_{i\to+\infty} x_{p_i}^+ = x_p^+ \text{ in } X.$$

(2) If $(-) \lim_{i\to+\infty} p_i = p$ and if $x_p^- \neq x_p^+$, then

$$\lim_{i\to+\infty} x_{p_i}^- = x_p^- = \lim_{i\to+\infty} x_{p_i}^+ \text{ in } X.$$

(3) If $(+) \lim_{i\to+\infty} p_i = p$ and if $x_p^- \neq x_p^+$, then

$$\lim_{i\to+\infty} x_{p_i}^+ = x_p^+ = \lim_{i\to+\infty} x_{p_i}^- \text{ in } X.$$

(4) If $\lim_{i\to+\infty} x_{p_i}^* = x_p^*$ in X, then $\lim_{i\to+\infty} p_i = p$.

(5) If $\lim_{i\to+\infty} x_{p_i}^* = x_p^*$ in X and if $x_p^* = x_p^- \neq x_p^+$, then $(-) \lim_{i\to+\infty} p_i = p$.

(6) If $\lim_{i\to+\infty} x_{p_i}^* = x_p^*$ in X and if $x_p^- \neq x_p^+ = x_p^*$, then $(+) \lim_{i\to+\infty} p_i = p$.

12.62. Remark. The topology of Γ induced by the topology of X coincides with $\mathfrak{J}$.

12.63. THEOREM. *The following statements are valid:*

(1) Γ *is a totally minimal orbit-closure under* σ.

(2) $\sigma \mid \Gamma$ *is locally almost periodic.*

(3) Γ *contains only one pair of nonseparated orbits, namely, the orbits of* $x_a^- -$ *and* $x_a^+ +$ *and* $x_a^- - \updownarrow x_a^+ -$.

(4) *If* φ *is a homeomorphism of* Γ *onto* Γ *such that* $\varphi\sigma = \sigma\varphi$, *then* $\varphi = \sigma^n \mid \Gamma$ *for some integer* n.

PROOF. (1) and (2) Cf. Hedlund [4].

(3) Obvious.

(4) Use 10.10.

12.64. NOTES AND REFERENCES.

References to the literature on symbolic dynamics up to 1940 can be found in the papers of Morse and Hedlund [3, 4].

(12.28) The bisequence μ was originally defined by Morse (cf. Morse [1]) and used by him to study the behavior of the geodesics on surfaces of negative curvature.

(12.31) The equality of μ and θ was noted by G. D. Birkhoff ([1], vol. 1, p. 691).

(12.47) The constructive method described for obtaining all the bisequences in the minimal set M is due to S. Kakutani (Personal communication).

(12.51) The bisequence η was defined by Garcia and Hedlund [1] without reference to the bisequence μ and the equivalence of their definition with (12.52) was observed by J. C. Oxtoby.

(12.57) The bisequence defined is Sturmian (cf. Hedlund [4]).

13. GEODESIC FLOWS OF MANIFOLDS OF CONSTANT NEGATIVE CURVATURE

13.01. THE HYPERBOLIC PLANE. Let U be the unit circle $z\bar{z} = x^2 + y^2 = 1$ of the complex z-plane $\mathcal{C}$ and let M be its interior. The *hyperbolic plane* $\mathfrak{M}$ is the two-dimensional analytic Riemannian manifold defined by assigning to M the differential metric

$$ds^2 = \frac{4(dx^2 + dy^2)}{(1 - x^2 - y^2)^2} = \frac{4\, dz\, d\bar{z}}{(1 - z\bar{z})^2}\,. \tag{1}$$

If φ is an arc of Class D^1 in $\mathfrak{M}$, the *hyperbolic length* of φ, or *h-length* of φ, denoted $\mathcal{L}_h(\varphi)$, is

$$\mathcal{L}_h(\varphi) = \int_{\operatorname{dom}\varphi} \frac{2\sqrt{\dot{x}^2 + \dot{y}^2}}{1 - x^2 - y^2}\, dt. \tag{2}$$

Since in (1) we have $g_{11} = g_{22}$ and $g_{12} = 0 = g_{21}$, angle in $\mathfrak{M}$ coincides with euclidean angle.

Let E be a measurable subset of $\mathfrak{M}$. The *hyperbolic area* of E, or *h-area* of E, denoted by $\mathcal{A}_h(E)$, is

$$\mathcal{A}_h(E) = \int_E \frac{4\, dx\, dy}{(1 - x^2 - y^2)^2}\,. \tag{3}$$

13.02. GEODESICS IN THE HYPERBOLIC PLANE. Let $\mathfrak{M}^*$ be the two-dimensional analytic Riemannian manifold defined by assigning the differential metric

$$ds^2 = \frac{dx^2 + dy^2}{y^2} \tag{4}$$

to the space $M^* = [z \mid z \in \mathcal{C}, \mathcal{J}(z) > 0]$. Let $\mathcal{S}$ denote the complex sphere. The transformation

$$\left(\frac{iz + i}{-z + 1}\,\middle|\, z \in \mathcal{S}\right) \tag{5}$$

is an analytic homeomorphism of $\mathcal{S}$ onto $\mathcal{S}$. If μ denotes the restriction of (5) to M, then μ is an analytic isometry of $\mathfrak{M}$ onto $\mathfrak{M}^*$.

The following three sets coincide:

(α) The set of all geodesics parameterized by arclength in $\mathfrak{M}^*$.

(β) The set of all curves $((x(s), y(s)) \mid s \in \mathcal{R})$ of Class C^2 in $\mathfrak{M}^*$ parameterized by arclength in $\mathfrak{M}^*$ such that

$$y(s)\, \ddot{x}(s) - 2\dot{x}(s)\dot{y}(s) = 0 \qquad (s \in \mathcal{R}),$$

$$y(s)\, \ddot{y}(s) - 2(\dot{y}(s))^2 + (y(s))^2 = 0 \qquad (s \in \mathcal{R}).$$

(γ) The union of the following four sets of curves:
$[((a \tanh (s + c) + b, a \operatorname{sech} (s + c)) \mid s \in \mathcal{R}) \mid a, b, c \in \mathcal{R} \ \& \ a > 0]$,
$[((a \tanh (-s + c) + b, a \operatorname{sech} (-s + c)) \mid s \in \mathcal{R}) \mid a, b, c \in \mathcal{R} \ \& \ a > 0]$,
$[((a, e^{s+b}) \mid s \in \mathcal{R}) \mid a, b \in \mathcal{R}]$,
$[((a, e^{-s+b}) \mid s \in \mathcal{R}) \mid a, b \in \mathcal{R}]$.

It follows that $\mathfrak{M}^*$ is complete (in the sense of Hopf and Rinow [2]) and the range of any geodesic in $\mathfrak{M}^*$ is the intersection of M with a circle in $\mathcal{S}$ which is orthogonal to the x-axis. Thus $\mathfrak{M}$ is complete and the range of a geodesic in $\mathfrak{M}$ is the intersection of M with a circle in $\mathcal{C}$ which is orthogonal to U.

13.03. Hyperbolic lines, rays and line segments. Let C be a circle in $\mathcal{S}$ which is orthogonal to U. The set $C \cap \mathfrak{M}$ is called a *hyperbolic line* or *h-line*. Any arc of C which together with its endpoints p and q lies in $\mathfrak{M}$ is called a *hyperbolic line segment* or *h-line segment* and is said to *join p and q*. Given two different points p and q of $\mathfrak{M}$, there exists a unique h-line segment joining p and q. Let L be an h-line and let $p \in L$. Either of the two components of $L - p$ together with p is called a *hyperbolic ray* or *h-ray* of which p is called the *initial point*.

Let L be an h-line. Then $L = C \cap \mathfrak{M}$, where C is a circle in $\mathcal{S}$ orthogonal to U. The two points in which C meets U will be called the *points at infinity* of L. Given different points u and v of U, there exists a unique h-line with u and v as its points at infinity.

Let R be an h-ray. The set R has just one limit point on U and this point is called the *point at infinity* of R. Given $p \in \mathfrak{M}$ and $u \in U$, there exists a unique h-ray with p as initial point and with u as point at infinity.

13.04. Hyperbolic distance. Let p and q be different points of $\mathfrak{M}$ and let S be the unique h-line segment with endpoints p and q. All geodesic arcs in $\mathfrak{M}$ with range S have the same h-length and this h-length, denoted by $D(p, q)$, will be called the *hyperbolic distance* or *h-distance between p and q*. The h-distance between p and q is the greatest lower bound of the h-lengths of curves of Class D^1 in $\mathfrak{M}$ and joining p and q (cf. Hopf and Rinow [2]).

13.05. Isometries of the hyperbolic plane. Let τ be an analytic isometry of $\mathfrak{M}$ onto $\mathfrak{M}$. Then τ is a conformal (directly or indirectly) analytic homeomorphism of M onto M and τ admits a unique extension τ^* to $M \cup U$ such that τ^* is a homeomorphism of $M \cup U$ onto $M \cup U$. Let τ^* have distinct fixed points $u, v \in U$. If h denotes the hyperbolic line with points at infinity u and v, then $h\tau = h$ and either all points of h are fixed under τ or no point of h has this property. In the latter case it is said that τ is an *isometry of* $\mathfrak{M}$ *onto* $\mathfrak{M}$ *with axis h* and h is called an *axis with endpoints u and v*. If τ is an isometry of $\mathfrak{M}$ onto $\mathfrak{M}$ with axis h and τ advances points of h toward v, v is the *positive fixed point* of τ and u is the *negative fixed point* of τ. The points u and v are the only fixed points of τ^* in $M \cup U$. If V is any neighborhood in $\mathcal{C}$ of $\{u\}\{v\}$ and A is any subset of $M \cup U$ such that $u, v \notin \overline{A}$, there exists an integer $N > 0$ such that $\{n < -N, n \in \mathfrak{g}\}\{n > N, n \in \mathfrak{g}\}$ implies $A(\tau^*)^n \subset V$.

To a large extent, consideration of isometries of $\mathfrak{M}$ onto $\mathfrak{M}$ with axes will suffice for later developments, but the collection of all isometries of $\mathfrak{M}$ onto $\mathfrak{M}$ admits a complete and simple analysis which we develop briefly.

Let Σ^+ denote the group of linear fractional transformations

$$((az + \bar{c})/(cz + \bar{a}) \mid z \in \mathcal{S})$$

where $a, c \in \mathcal{C}$ with $a\bar{a} - c\bar{c} = 1$. Let $\sigma \in \Sigma^+$. Then $U\sigma = U$, $M\sigma = M$, σ is a directly conformal analytic homeomorphism of $\mathcal{S}$ onto $\mathcal{S}$ and any transformation with these properties is an element of Σ^+. It is easily verified that the restriction of σ to M is an analytic isometry of $\mathfrak{M}$ onto $\mathfrak{M}$.

Let $\sigma = ((az + \bar{c})/(cz + \bar{a}) \mid z \in \mathcal{S}) \in \Sigma^+$ and suppose σ is not the identity. Since $a + \bar{a}$ is real, σ must be either hyperbolic with fixed points on U, parabolic with fixed point on U, or elliptic with fixed points inverse to U.

Let $\sigma \in \Sigma^+$, let σ be hyperbolic, and let $u, v \in U$ be the fixed points of σ. Let $p \in \mathcal{S}$ with $p \neq u, v$. Then the sequence $p, p\sigma^{-1}, p\sigma^{-2}, \cdots$ converges to one of the fixed points of σ and the sequence $p, p\sigma, p\sigma^2, \cdots$ converges to the other fixed point of σ. The first of these points will be called the *negative fixed point* of σ, and the other will be called the *positive fixed point* of σ. Let u^- be the negative fixed point of σ, let u^+ be the positive fixed point of σ and let A be any subset of $\mathcal{S}$ such that $u^-, u^+ \notin \bar{A}$. Then if V is a neighborhood of $\{u^-\}\{u^+\}$ there exists a positive integer N such that $\{n < -N, n \in \mathcal{I}\}\{n > N, n \in \mathcal{I}\}$ implies $A\sigma^n \subset V$. Any circular arc of $\mathcal{S}$ with endpoints u^- and u^+ is invariant under σ. The *axis* of σ is the hyperbolic line h with points at infinity u^-, u^+, and these points will also be called the *endpoints* of the axis. Let $\tau = \sigma \mid M$. Then τ is an isometry of $\mathfrak{M}$ onto $\mathfrak{M}$ with $\{$negative$\}\{$positive$\}$ fixed point $\{u^-\}\{u^+\}$ and with axis h.

Let $\sigma \in \Sigma^+$, let σ be parabolic, and let $u \in U$ be the fixed point of σ. If A is any subset of $\mathcal{S}$ such that $u \notin \bar{A}$, and V is any neighborhood of u, then there exists an integer $N > 0$ such that $|n| > N$, $n \in \mathcal{I}$, implies $A\sigma^n \subset V$. If C is any circle which is tangent to U at u then $C\sigma = C$.

Let $\sigma \in \Sigma^+$ and let σ be elliptic. Then σ has two fixed points in $\mathcal{S}$ which are inverse with respect to U. There exists a disjoint class of circles covering $\mathcal{S}$ except for the fixed points of σ such that each member of this class is invariant under σ and the fixed points of σ are inverse with respect to each member of this class.

Let Σ^- denote the set of transformations $((a\bar{z} + \bar{c})/(c\bar{z} + \bar{a}) \mid z \in \mathcal{S})$ where $a, c \in \mathcal{C}$ with $a\bar{a} - c\bar{c} = 1$. Let $\sigma \in \Sigma^-$. Then $U\sigma = U$, $M\sigma = M$, σ is an inversely conformal analytic homeomorphism of $\mathcal{S}$ onto $\mathcal{S}$ and any transformation with these properties is a member of Σ^-. The restriction of σ to M is an analytic isometry of $\mathfrak{M}$ onto $\mathfrak{M}$.

Let $\sigma \in \Sigma^-$. Then, either there exists a circle $C \subset \mathcal{S}$ orthogonal to U such that $p \in C$ implies $p\sigma = p$, or σ has exactly two fixed points u, v and $u, v \in U$. In the first case σ is an inversion in C. In the second case, let τ be the inversion in the circle D which passes through u and v and which is orthogonal to U. Let

$\sigma_1 = \sigma\tau$. Then σ_1 is directly conformal, $U\sigma_1 = U$, $M\sigma_1 = M$, σ_1 cannot be the identity mapping, $\sigma_1 \in \Sigma^+$ and $u\sigma_1 = u$, $v\sigma_1 = v$. Thus σ_1 is a hyperbolic transformation with fixed points u and v. We have $\sigma = \tau^{-1}\sigma_1 = \tau\sigma_1 = \sigma_1\tau$ and we conclude that σ is the product of a hyperbolic transformation and an inversion in a circle orthogonal to U which passes through the fixed points of σ. The transformation σ is called a *paddle motion*. The {*negative*}{*positive*} fixed point of σ is defined to be the {negative}{positive} fixed point of σ_1. If A is any subset of $\mathcal{S}$ such that $u, v \notin \bar{A}$ and V is any neighborhood of the {negative}{positive} fixed point of σ, there exists an integer $N > 0$ such that $\{n < -N, n \in \mathfrak{g}\}$ $\{n > N, n \in \mathfrak{g}\}$ implies $A\sigma^n \subset V$. The *axis* of σ is the axis of σ_1. Let $\tau = \sigma \mid M$. Then τ is an isometry of $\mathfrak{M}$ onto $\mathfrak{M}$ with axis h and with {negative}{positive} fixed point that of σ.

Let $\Sigma = \Sigma^- \cup \Sigma^+$. Then Σ is a group of conformal analytic homeomorphisms of $\mathcal{S}$ onto $\mathcal{S}$ and any conformal analytic homeomorphism τ of $\mathcal{S}$ onto $\mathcal{S}$ such that $U\tau = U$ and $M\tau = M$ is an element of Σ. Let $\tau \in \Sigma$ and let $\sigma = \tau \mid M$. Then σ is an analytic isometry of $\mathfrak{M}$ onto $\mathfrak{M}$, and conversely, if σ is an analytic isometry of $\mathfrak{M}$ onto $\mathfrak{M}$, then there exists a unique element $\tau \in \Sigma$ such that $\sigma = \tau \mid M$. This extension of σ will be denoted by $\hat{\sigma}$.

13.06. Hyperbolic circles and equidistant curves. Since the differential metric 13.01(1) is invariant under rotations about the origin, the locus of points of $\mathfrak{M}$ at constant h-distance from the origin is a circle with center at the origin O. If p is any point of $\mathfrak{M}$, there exists $\sigma \in \Sigma$ such that $O\sigma = p$; thus the locus of points at constant h-distance r from p is again a circle C containing p in its interior, though unless p is at the origin the euclidean center of C will not coincide with p. We call C the *hyperbolic circle* or *h-circle with center p and with radius r*. It is invariant under the group of elliptic transformations which have p and its inverse in U as fixed points.

Given an h-line L which does not pass through the origin O, there is a unique h-line passing through O and orthogonal to L. By application of a suitable transformation belonging to Σ we see that the same is true for any h-line L and point $p \in \mathfrak{M}$, p not on L. Let q be the intersection of L and the h-line through p orthogonal to L. The h-line segment pq is the *hyperbolic perpendicular from p to L* and $D(p, q)$ is the *h-distance from p to L*. The h-distance from p to L is denoted $D(p, L)$. The point q is the *foot of the perpendicular* from p to L. By consideration of the case where p is at the origin, it is evident that the h-distance from p to L is less than the h-length of any h-line segment joining p to any point of L other than the foot of the perpendicular from p to L.

Let L be an h-line with points at infinity u and v. Let C be a circle passing through u and v with $C \neq U$ and C not orthogonal to U. Let $A = C \cap M$. Then A is invariant under every hyperbolic transformation with u and v as fixed points. Let $p, q \in A$. There exists a linear fractional transformation σ such that $u\sigma = u$, $v\sigma = v$, $p\sigma = q$. But then $M\sigma = M$, $U\sigma = U$, $\sigma \in \Sigma^+$, $A\sigma = A$, and σ is a hyperbolic transformation with axis L. The hyperbolic perpendicular from p to L is transformed by σ into the hyperbolic perpendicular from q to

L and thus all points of A are at the same h-distance d from L. We call d the *h-distance from A to L.*

13.07. HOROCYCLES. Let C be a euclidean circle which is internally tangent to U at $u \in U$ and let $H = C - u$. Then H is invariant under any parabolic transformation in Σ^+ which has u as fixed point and H is an orthogonal trajectory of the family of h-lines which have u as common point at infinity. We call H a *horocycle* and u its *point at infinity.*

Let H be a horocycle and let A be a circular arc such that $A \subset H$. We can parameterize A with euclidean arclength and thus define an arc in $\mathfrak{M}$ with A as its range; the h-length of this arc will be called the h-length of A.

Let L_1 and L_2 be h-lines with common point at infinity u and let H_1 and H_2 be horocycles with u as common point at infinity and such that H_1 is interior to $H_2 \cup u$. Let $\{s_1\}\{s_2\}$ denote the h-length of the arc of $\{H_1\}\{H_2\}$ cut off by L_1 and L_2 and let s be the h-length of the h-line segment of L_1 (or L_2) cut off by H_1 and H_2. Then

$$s_2 = s_1 e^s. \tag{6}$$

To derive this formula we can assume that u is any point of U and, in particular, we can choose $u = +1$. Under the transformation 13.02(5), which is an analytic isometry of $\mathfrak{M}$ onto $\mathfrak{M}^*$, the image of $\{L_1\}\{L_2\}$ is the range of a geodesic $\{(x(t) = a_1, y(t) = e^t) \mid t \in R\}\{(x(t) = a_2, y(t) = e^t) \mid t \in R\}$ and the image of $\{H_1\}$ $\{H_2\}$ is the range of the curve

$$\{(x(t) = t, y(t) = d_1) \mid \min(a_1, a_2) \leqq t \leqq \max(a_1, a_2)\}$$

$$\{(x(t) = t, y(t) = d_2) \mid \min(a_1, a_2) \leqq t \leqq \max(a_1, a_2)\} \text{ with } d_1 > d_2 .$$

We then have

$$s_i = \left| \int_{a_1}^{a_2} d_i^{-1}\, dt \right| = d_i^{-1} \mid a_1 - a_2 \mid, \qquad i = 1, 2,$$

and

$$s = \int_{d_2}^{d_1} \frac{dy}{y} = \log (d_1/d_2)$$

which imply (6).

13.08. ASYMPTOTIC GEODESICS IN $\mathfrak{M}$. Let φ be a geodesic parametrized by arclength in $\mathfrak{M}$ and let L be the range of φ. The sequence $\{\varphi(-1), \varphi(-2), \cdots\}$ $\{\varphi(1), \varphi(2), \cdots\}$ converges in $\mathfrak{C}$ to a point $\{u^- \in U\}\{u^+ \in U\}$. The points u^-, u^+ are the points at infinity of L. The point $\{u^-\}\{u^+\}$ is called the $\{$*negative*$\}\{$*positive*$\}$ *point at infinity* of φ. If $\{V^-\}\{V^+\}$ is an open subset of $\mathfrak{C}$ containing $\{u^-\}\{u^+\}$ then there exists $\{s^- \in \mathfrak{R}\}\{s^+ \in \mathfrak{R}\}$ such that $\{s \leqq s^-\}$ $\{s \geqq s^+\}$ implies $\{\varphi(s) \in V^-\}\{\varphi(s) \in V^+\}$.

Let φ and ψ be geodesics parameterized by arclength in $\mathfrak{M}$ let $\{u^-, u^+\}$

$\{v^-, v^+\}$ be the negative, positive points at infinity of $\{\varphi\}\{\psi\}$ and let $L = \operatorname{rng} \psi$. Then

(a) $u^- = v^-$ and $u^+ \neq v^+$ implies

$$\lim_{r \to -\infty} D(\varphi(r), L) = 0, \qquad \lim_{r \to +\infty} D(\varphi(r), L) = +\infty.$$

(b) $u^- \neq v^-$ and $u^+ = v^+$ implies

$$\lim_{r \to -\infty} D(\varphi(r), L) = +\infty, \qquad \lim_{r \to +\infty} D(\varphi(r), L) = 0.$$

(c) $u^- \neq v^-$ and $u^+ \neq v^+$ implies

$$\lim_{r \to -\infty} D(\varphi(r), L) = +\infty = \lim_{r \to +\infty} D(\varphi(r), L).$$

Let φ and ψ be geodesics parameterized by arclength in $\mathfrak{M}$ with different ranges. Then φ and ψ are said to be $\{$*negatively*$\}\{$*positively*$\}$ *asymptotic* provided there exists $s_0 \in \mathfrak{R}$ such that $\{\lim_{s \to -\infty} D(\varphi(s), \psi(s + s_0)) = 0\}\{\lim_{s \to +\infty} D(\varphi(s), \psi(s + s_0)) = 0\}$, or equivalently, provided there exists $s_1 \in \mathfrak{R}$ such that $\{\lim_{s \to -\infty} D(\varphi(s + s_1), \psi(s)) = 0\}\{\lim_{s \to +\infty} D(\varphi(s + s_1), \psi(s)) = 0\}$.

13.09. THEOREM. *Let φ and ψ be geodesics parameterized by arclength in $\mathfrak{M}$ with different ranges. Then φ and ψ are $\{$negatively$\}\{$positively$\}$ asymptotic if and only if φ and ψ have the same $\{$negative$\}\{$positive$\}$ point at infinity.*

PROOF. Let $\{u^-, u^+\}\{v^-, v^+\}$ be the negative, positive points at infinity of $\{\varphi\}\{\psi\}$. Suppose that $u^- \neq v^-$. Let $d(s) = D(\varphi(s), \operatorname{rng} \psi)$. Then

$$\lim_{s \to -\infty} d(s) = +\infty.$$

Let $s_0 \in \mathfrak{R}$. Then $D(\varphi(s), \psi(s + s_0)) \geqq d(s)$ and thus $\lim_{s \to -\infty} D(\varphi(s), \psi(s + s_0)) = +\infty$. It follows that φ and ψ are not negatively asymptotic.

Similarly, if $u^+ \neq v^+$ then φ and ψ are not positively asymptotic.

Now suppose $u^- = v^-$. Let H be the horocycle with u^- as point at infinity and such that $\varphi(0) \in H$. Let $\psi(s_0)$ be the point in which H meets the range of ψ and let t be the h-length of the arc of H with endpoints $\varphi(0)$ and $\psi(s_0)$. From (6) of 13.07 we have

$$D(\varphi(s), \psi(s + s_0)) \leqq te^s, \qquad s \in R.$$

Thus $\lim_{s \to -\infty} D(\varphi(s), \psi(s + s_0)) = 0$ and φ and ψ are negatively asymptotic.

Similarly, if $u^+ = v^+$ then φ and ψ are positively asymptotic.

The proof is completed.

13.10. THE GEODESIC FLOW OVER $\mathfrak{M}$. Let $p \in \mathfrak{M}$. *A unitangent on $\mathfrak{M}$ at p* is a unit contravariant vector at p. The *unitangent space on $\mathfrak{M}$ at p*, denoted $\mathfrak{J}(\mathfrak{M}, p)$ is the set of all unitangents on $\mathfrak{M}$ at p.

The *unitangent space on* $\mathfrak{M}$, denoted X, is $\bigcup_{p\in\mathfrak{M}} \mathfrak{J}(\mathfrak{M}, p)$. Let the transformation μ of X onto $\mathfrak{M}$ be defined as follows. If $x \in X$ and x is a unitangent at $p \in \mathfrak{M}$, then $x\mu = p$. The transformation μ is the *projection of* X *onto* $\mathfrak{M}$.

Let $x_1, x_2 \in X$, let $p_1 = x_1\mu$, $p_2 = x_2\mu$ and let $\delta(x_1, x_2)$ denote the absolute value of the angle between x_2 and the unitangent obtained by parallel displacement of x_1 to p_2 along the unique geodesic segment joining p_1 and p_2. For $r \in \mathfrak{R}^+$, define $\alpha_r = [(x_1, x_2) \mid x_1, x_2 \in X, h(x_1\mu, x_2\mu) + \delta(x_1, x_2) < r]$. Define $\mathfrak{V} = [\alpha_r \mid r \in \mathfrak{R}]$. It is readily verified that $\mathfrak{V}$ is a uniformity-base of X. Let $\mathfrak{U}$ be the uniformity generated by $\mathfrak{V}$. We provide X with this uniformity and assign to X the topology induced by $\mathfrak{U}$.

Let $G = \Sigma \mid M$ and let $g \in G$. Then g is an isometry of $\mathfrak{M}$ onto $\mathfrak{M}$, g defines a homeomorphism dg of X onto X and the set $dG = [dg \mid g \in G]$ is a homeomorphism group of X which is universally transitive, and dG is uniformly equicontinuous relative to $\mathfrak{U}$. If α is an open index of X, there exists a nonvacuous open subset E of X such that for $(x, y) \in X$ we have $(x, y) \in \alpha$ if and only if $x\, dg, y\, dg \in E$ for some $g \in G$.

Let γ be the transformation of $X \times \mathfrak{R}$ onto X defined as follows. Let $x \in X$ and let $s \in \mathfrak{R}$. Let φ be a geodesic parameterized by arclength in $\mathfrak{M}$ such that $x = \dot{\varphi}(s_0)$ is the tangent vector to φ at $\varphi(s_0)$. Let $y = \dot{\varphi}(s + s_0)$. We define $(x, s)\gamma = y$. Then $\mathcal{G} = (X, \mathfrak{R}, \gamma)$ is a transformation group of X which is called the *geodesic flow of* $\mathfrak{M}$.

Let φ be a geodesic parameterized by arclength in $\mathfrak{M}$. The set of tangent vectors to φ at all elements of $\mathfrak{R}$ is an orbit under $\mathcal{G}$ and is denoted by O_φ.

13.11. THEOREM. *Let φ and ψ be geodesics parameterized by arclength in $\mathfrak{M}$ with different ranges. Then O_φ and O_ψ are* {*negatively*} {*positively*} *asymptotic if and only if φ and ψ have the same* {*negative*} {*positive*} *point at infinity.*

PROOF. The necessity follows from 13.09 and the fact that the projection $\mu : X \to \mathfrak{M}$ of X onto $\mathfrak{M}$ is uniformly continuous.

To prove the sufficiency, let φ and ψ have the same positive point at infinity u. (Proof of the other case is similar.) We may suppose there exists a horocycle $H(0)$ with u as point at infinity and such that $\varphi(0), \psi(0) \in H(0)$. Let $H(s)$ $(s \in \mathfrak{R})$ be the horocycle with u as point at infinity and such that $\varphi(s), \psi(s) \in H(s)$. Let v be the negative point at infinity of φ. Let $\sigma(s)$ $(s \in \mathfrak{R})$ be the hyperbolic transformation whose axis has u, v as endpoints and such that $\varphi(s)\sigma(s) = \varphi(0)$. Then $H(s)\sigma(s) = H(0)$ and the unitangents $\dot{\varphi}(s)\, d\sigma(s) = \dot{\varphi}(0)$, $\dot{\psi}(s)\, d\sigma(s)$ are unitangents at $\varphi(0), \psi(s)\sigma(s) \in H(0)$ which are internally orthogonal to $H(0)$. Since $\sigma(s)$ is an isometry of $\mathfrak{M}$ onto $\mathfrak{M}$, $D(\varphi(0), \psi(s)\sigma(s)) \leqq h(s) = e^{-s}h(0)$, where $h(s)$ is the arclength of the arc of $H(s)$ with endpoints $\varphi(s), \psi(s)$. The conclusion follows.

It is now clear that if φ and ψ are geodesics parameterized by arclength in $\mathfrak{M}$ with different ranges such that O_φ and O_ψ are neither negatively nor positively asymptotic, then there exists a geodesic θ parameterized by arclength in $\mathfrak{M}$ such that O_θ is negatively asymptotic to O_φ and positively asymptotic to O_ψ.

13.12. THE HOROCYCLE FLOW. For $x \in X$ let φ_x be the geodesic parameterized by arclength in $\mathfrak{M}$ such that $x = \dot{\varphi}_x(0)$, let $u^-(x)$ be the negative point at infinity of φ_x and let $\eta_x : \mathfrak{R} \to \mathfrak{M}$ be the analytic curve parameterized by arclength in $\mathfrak{M}$ such that rng η_x is a horocycle with u^- as point at infinity, such that $\eta_x(0) = x\mu$, and such that $(\dot{\eta}_x(0), x)$ is positively oriented. Define $\kappa : X \times \mathfrak{R} \to X$ such that $x \in X$, $s \in \mathfrak{R}$ implies that $(x, s)\kappa$ is the unitangent at $\eta_x(s)$ which is externally orthogonal to rng η_x. The continuous flow $(X, \mathfrak{R}, \kappa)$ is called the *horocycle flow of* $\mathfrak{M}$ and denoted by $\mathfrak{H}$.

13.13. LIMIT SET OF A SUBGROUP OF Σ. Let Ω denote a subgroup of Σ. If $p, q \in M$, then $\overline{p\Omega} \cap U = \overline{q\Omega} \cap U$. The *limit set of* Ω, denoted $\Lambda(\Omega)$, is the set $\overline{p\Omega} \cap U$ where $p \in M$.

Clearly $\Lambda(\Omega)$ is closed and invariant under Ω.

13.14. LEMMA. *Let E be a finite subset of U, let $u \in \Lambda(\Omega)$ and let A be an arc of U with midpoint u. Then there exists $\omega \in \Omega$ such that $E\omega \cap (U - A)$ consists of at most one point.*

PROOF. Let β be the least angle formed at the origin O by pairs of h-rays with initial point O and with points at infinity distinct points of E. Since E is finite, $\beta > 0$. Let $p \in M$ and let $\alpha(p)$ be the angle subtended by A at p by h-rays with initial point p and points at infinity the endpoints of A. Then $\lim_{p\to u} \alpha(p) = 2\pi$. There exists a sequence $\omega_1, \omega_2, \cdots$ of elements of Ω such that $\lim_{n\to\infty} O\omega_n = u$. Let N be a positive integer such that $\alpha(O\omega_N) > 2\pi - \beta$. Since ω_n is conformal, $(U - A)\omega_N^{-1}$ can contain at most one point of E. It follows that $E\omega_N \cap (U - A)$ consists of at most one point. The proof is completed.

13.15. THEOREM. *If* crd $\Lambda(\Omega)$ *is finite, then* crd $\Lambda(\Omega) = 0, 1$ *or* 2. *If* crd $\Lambda(\Omega)$ *is not finite, then $\Lambda(\Omega)$ is self-dense and either $\Lambda(\Omega) = U$ or $\Lambda(\Omega)$ is nowhere dense on U.*

PROOF. We suppose that crd $\Lambda(\Omega) > 2$. Let a, b, c be different points of $\Lambda(\Omega)$. Let $u \in \Lambda(\Omega)$ and let A be an arc of U with midpoint u. Since $\omega \in \Omega$ implies that $a\omega, b\omega, c\omega \in \Lambda(\Omega)$, it follows from 13.14 that there exist at least two points of $\Lambda(\Omega)$ in A. Thus $\Lambda(\Omega)$ is self-dense.

To complete the proof it is sufficient to show that if $\Lambda(\Omega)$ contains an arc of U, then $\Lambda(\Omega) = U$. Assuming that the arc $A \subset \Lambda(\Omega)$, we let a be its midpoint and let $\delta \in R^+$. There exists $\omega \in \Omega$ such that the angle subtended by A at $O\omega$ by hyperbolic rays exceeds $2\pi - \delta$. But then the angle subtended by $A\omega^{-1}$ at O by hyperbolic rays also exceeds $2\pi - \delta$. Since $A\omega^{-1} \subset \Lambda(\Omega)$ and δ can be chosen arbitrarily small, we infer that $\Lambda(\Omega) = U$. The proof is completed.

13.16. THEOREM. *Let $u \in \Lambda(\Omega)$ and let* crd $u\Omega \geqq 2$. *Then* $\overline{u\Omega} = \Lambda(\Omega)$.

PROOF. Use 13.14.

13.17. THEOREM. *Let* crd $\Lambda(\Omega) > 2$. *Then Ω is transitive on $\Lambda(\Omega)$.*

PROOF. Let $a, b \in \Lambda(\Omega)$ and let A, B be open arcs of U with midpoints a, b. By 13.15, $\Lambda(\Omega)$ is self-dense, and thus there exist different points u, v such that $u, v \in \Lambda(\Omega) \cap A$. According to 13.14 there exists $\omega \in \Omega$ such that either $u\omega \in B$ or $v\omega \in B$. Since $u\omega, v\omega \in \Lambda(\Omega)$, the proof of the theorem is completed.

13.18. THEOREM. *Let* crd $\Lambda(\Omega) > 2$. *Then there exists at most one point of* $\Lambda(\Omega)$ *which is not transitive under* Ω.

PROOF. Let $u, v \in \Lambda(\Omega)$ and suppose that neither u nor v is transitive under Ω. It follows from 13.16 that crd $u\Omega = 1 =$ crd $v\Omega$, or, equivalently, that $u\Omega = u$, $v\Omega = v$, and thus, if L denotes the h-line with points at infinity u, v, then $L\omega = L$ for every $\omega \in \Omega$. Let $p \in L$. It is evident that $U \cap \overline{p\Omega}$ can contain at most the two points u and v, and thus crd $\Lambda(\Omega) \leqq 2$, contrary to hypothesis. The proof is completed.

13.19. REMARK. Let $\sigma \in \Sigma$ and let there exist a closed interval A of U such that $A\sigma \subset$ int A. Then σ has an axis with the positive fixed point of σ in int A and the negative fixed point of σ in int $(U - A)$.

13.20. THEOREM. *Let* crd $\Lambda(\Omega) > 2$. *Then some element of* Ω *has an axis.*

PROOF. Suppose that no element of Ω has an axis. Let A be a closed interval of U such that $A \neq U$ and let $A^* = U -$ int A. It follows from 13.19 that if $\omega \in \Omega$, then $A^*\omega \cap A \neq \emptyset$.

We show that if $u \in \Lambda(\Omega)$ and B is a closed interval of U such that $u \notin B$, then there exists $\omega \in \Omega$ such that $B\omega \cap B = \emptyset$. Let B be such an interval and let C be a closed interval of U such that $u \in$ int C and $B \cap C = \emptyset$. Let D be a closed interval of U such that $C \subset$ int D and $D \cap B = \emptyset$. Let α be a positive number less than either of the two angles subtended by h-rays at the origin O by the two intervals which constitute $D - C$. There exists $\zeta \in \Omega$ such that the angle which $C' = U - C$ subtends by h-rays at $O\zeta$ is less than α. But then the angle which $C'\zeta^{-1}$ subtends by h-rays at O is less than α. Let $C^* = U -$ int C. Then the angle which $C^*\zeta^{-1}$ subtends by h-rays at O is also less than α and since $C^*\zeta^{-1} \cap C \neq \emptyset$, we infer that $C^*\zeta^{-1} \subset D$. Let $\omega = \zeta^{-1}$. Then $B\omega \subset C^*\zeta^{-1} \subset D$ and $B\omega \cap B = \emptyset$.

Now let u and v be distinct points of $\Lambda(\Omega)$. Let A, B be open disjoint intervals of $\Lambda(\Omega)$ containing u, v respectively. It has been shown that there exists $\omega_1 \in \Omega$ such that $A'\omega_1 \subset A$ and $\omega_2 \subset \Omega$ such that $B'\omega_2 \subset B$. But then $A'\omega_1\omega_2 \subset B \subset$ int A', contrary to the assumption that no element of Ω has an axis. The proof is completed.

13.21. LEMMA. *Let* crd $\Lambda(\Omega) > 2$. *Then there exist infinitely many distinct axes of transformations of* Ω *and the set of endpoints of these axes is dense in* $\Lambda(\Omega)$.

PROOF. It follows from 13.20 that there exists $\omega \in \Omega$ such that ω has an axis. Let L be the axis of ω and let u, v be the endpoints of L. Let $\sigma \in \Omega$. Then $L\sigma$ is an axis of $\sigma^{-1}\omega\sigma$, which is an element of Ω, and the endpoints of $L\sigma$ are $u\sigma$, $v\sigma$. Now use 13.18.

13.22. DEFINITION. The subgroup Ω of Σ is *mobile* provided that no point of U is fixed under all elements of Ω.

13.23. LEMMA. *Let Ω be a mobile subgroup of Σ and let* crd $\Lambda(\Omega) > 2$. *Then there exists a pair of axes of transformations of Ω such that these axes have no endpoints in common.*

PROOF. Let L, with endpoints a and b, be an axis of $\omega \in \Omega$. Not all of the infinitely many distinct axes of transformations of Ω have a as common endpoint. For if this were the case, since some element $\sigma \in \Omega$ moves a, there would be infinitely many distinct axes with $a\sigma$ as endpoint, of which some one would not have a as endpoint. Thus there exists an axis L_1 of $\omega_1 \in \Omega$ which does not have a as endpoint. Similarly, there exists an axis L_2 of $\omega_2 \in \Omega$ which does not have b as endpoint.

If the statement of the lemma is not true, L_1 must have b as one of its endpoints, L_2 must have a as one of its endpoints, and L_1 and L_2 must have a common endpoint c. But ω_1 moves L_2 into an axis which has no common endpoint with L.

The proof of the lemma is completed.

13.24. THEOREM. *Let Ω be a mobile subgroup of Σ with* crd $\Lambda(\Omega) > 2$. *Then*

(1) *$\Lambda(\Omega)$ is minimal under Ω.*

(2) *If A and B are open arcs of U such that $A \cap \Lambda(\Omega) \neq \emptyset \neq B \cap \Lambda(\Omega)$, then there exists $\omega \in \Omega$ such that ω has an axis L with endpoints a, b such that $a \in A$ and $b \in B$.*

(3) *If Φ is the set of all geodesics φ parameterized by arclength in M such that the points at infinity of φ both belong to $\Lambda(\Omega)$ and if Ψ is the set of all geodesics ψ parameterized by arclength in M such that* rng ψ *is an axis of some member of Ω, then $\bigcup_{\psi \in \Psi} O_\psi$ is dense in $\bigcup_{\varphi \in \Phi} O_\varphi$.*

(4) *The space product of the transformation group $(\Lambda(\Omega), \Omega \mid \Lambda(\Omega))$ with itself is transitive.*

PROOF. (1) Use 13.16.

(2) We can assume that $\bar{A} \cap \bar{B} = \emptyset$. Let $u \in A \cap \Lambda(\Omega)$ and let $v \in B \cap \Lambda(\Omega)$. It follows from 13.23 that there exist ω_1, $\omega_2 \in \Omega$ such that ω_1 has an axis L_1, ω_2 has axis L_2, and L_1 and L_2 have no endpoint in common. Since, if $\sigma \in \Omega$ has axis L and $\omega \in \Omega$, then $L\omega$ is the axis of $\omega^{-1}\sigma\omega$, we can assume, in view of (1), that one endpoint of L_1 is in A. But then some power of ω_1 transforms L_2 into an axis L_a of ω_a such that both endpoints of L_a are in A. Similarly there exists $\omega_b \in \Omega$ with axis L_b such that both endpoints of L_b are in B.

Let $A' = U - A$ and let $B' = U - B$. There exists an integer n such that $A'\omega_a^n \subset A$, $A'\omega_a^{-n} \subset A$, $B'\omega_b^n \subset B$ and $B'\omega_b^{-n} \subset B$. Define $\omega = \omega_b^n\omega_a^n$. Then $\bar{A}\omega = \bar{A}\omega_b^n\omega_a^n \subset B\omega_a^n \subset A$ and $\bar{B}\omega^{-1} = \bar{B}\omega_a^{-n}\omega_b^{-n} \subset A\omega_b^{-n} \subset B$. It follows that ω is a transformation with the desired properties.

(3) Use (2).

(4) Let A, B, C, D be open arcs of U such that each intersects $\Lambda(\Omega)$. Choose $\omega_1 \in \Omega$ such that ω_1 has axis L_1 with endpoints $a \in A$ and $d \in D$. There exists an integer n such that $F = (B \cap \Lambda(\Omega))\omega_1^n \cap D \neq \emptyset$. Choose $\omega_2 \in \Omega$ such that

ω_2 has axis L_2 with endpoints $f \in F$ and $c \in C$. There exists an integer m such that $(A \cap \Lambda(\Omega))\omega_1^n\omega_2^m \cap C \neq \emptyset$. Then also $(B \cap \Lambda(\Omega))\omega_1^n\omega_2^m \cap D \neq \emptyset$.

The proof is completed.

13.25. GEODESIC PARTITION FLOWS OF $\mathfrak{M}$. Let Ω be a subgroup of Σ and let $G = \Omega \mid M$. Then G is a group of isometries of $\mathfrak{M}$. We observe that dG is a uniformly continuous homeomorphism group of X and $\mathcal{G}$ is dG motion preserving. Define the partition $X_G = [x\, dG \mid x \in X]$ and let X_G be provided with its partition uniformity as defined in 2.34. We observe that the partition X_G is star-indexed and thus, by 2.36, the projection of X onto X_G is uniformly continuous and uniformly open. Let $\mathcal{G}_G = [X_G, R, \gamma_G]$, called the *geodesic partition flow induced by G*, be the partition flow on X_G induced by $\mathcal{G}$. For $x \in X$ let φ_x be the unique geodesic parameterized by arclength in $\mathfrak{M}$ such that $\dot{\varphi}_x(0) = x$. We remark that if $x \in p \in X_G$, then p is periodic under $\mathcal{G}_G$ if and only if rng φ_x is the axis of some member of G.

13.26. DEFINITION. A subgroup Ω of Σ is $\{$*limit-partial*$\}\{$*limit-entire*$\}$ provided that $\{$crd $\Lambda(\Omega) = \aleph$ and $\Lambda(\Omega) \neq U\}\{\Lambda(\Omega) = U\}$.

We observe that if Ω is limit-partial, then $\Lambda(\Omega)$ is a Cantor discontinuum.

13.27. THEOREM. *Let Ω be a limit-entire, mobile subgroup of Σ and let $G = \Omega \mid M$. Then*

(1) *The set of all $\mathcal{G}_G$-periodic points of X_G is dense in X_G.*

(2) *$\mathcal{G}_G$ is regionally transitive.*

PROOF. Use 13.24.

COROLLARY. *Let Ω be a limit-entire, mobile subgroup of Σ and let $G = \Omega \mid M$. Then there exists a point of X_G which is transitive under the geodesic partition flow $\mathcal{G}_G$.*

PROOF. It is easily proved that the space X of unitangents on $\mathfrak{M}$ is a second-countable, locally compact, Hausdorff space, and it follows that X_G is also second-countable. Let θ be the projection of X onto X_G. Let $x \in X$, let A be a compact neighborhood of x in X and let $\{\mathcal{B}_n \mid n \in \mathcal{I}^+\}$ be a base for the open sets in X_G. Since the geodesic partition flow $\mathcal{G}_G$ is transitive there exists $x_1 \in$ Int A such that the orbit of $x_1\theta$ under $\mathcal{G}_G$ meets $\mathcal{B}_1$. There exists a compact neighborhood N_1 of x_1 in X with $N_1 \subset A$ and such that $y \in N_1$ implies that the orbit of $y\theta$ under $\mathcal{G}_G$ meets $\mathcal{B}_1$. There exists $x_2 \in$ Int N_1 such that the orbit of $x_2\theta$ under $\mathcal{G}_G$ meets $\mathcal{B}_2$ and thus there exists a compact neighborhood N_2 of x_2 in X with $N_2 \subset N_1$ and such that $y \in N_2$ implies that the orbit of $y\theta$ under $\mathcal{G}_G$ meets $\mathcal{B}_2$. Proceeding inductively, we define a sequence $N_1, N_2, \cdots$ of compact neighborhoods in X such that $N_1 \supset N_2 \supset \cdots$ and $y \in N_n$ implies that the orbit of $y\theta$ under $\mathcal{G}_G$ meets $\mathcal{B}_n$. But then $\bigcap_{i=1}^{\infty} N_i \neq \emptyset$ and $x^* \in \bigcap_{i=1}^{\infty} N_i$ implies that $x^*\theta$ is a transitive point of X_G under $\mathcal{G}_G$.

The proof is completed.

13.28. DEFINITION. Let (X, T, π) be a transformation group and let $T = \mathcal{R}$

or $\mathcal{I}$. The point $x \in X$ is {*positively*}{*negatively*} *transient under* T provided that, C being any compact subset of X, there exists $t_0 \in T$ such that $\{t > t_0\}$ $\{t < t_0\}$ implies $xt \notin C$. The orbit Φ under T is {*positively*}{*negatively*} *transient* provided there exists $x \in \Phi$ such that x is {positively}{negatively} transient.

13.29. THEOREM. *Let Ω be a limit partial subgroup of Σ and let $G = \Omega \mid M$. Let φ be a geodesic parameterized by arclength in $\mathfrak{M}$ such that the {negative}{positive} point at infinity of φ does not belong to $\Lambda(\Omega)$. Let θ be the projection of X onto X_G. Then $O_\varphi \theta$ is a {negatively}{positively} transient orbit under $\mathcal{G}_G$.*

PROOF. Let $\mathfrak{A}$ be a compact subset of X_G. There exists a compact subset A of X such that $A\theta \supset \mathfrak{A}$. Let the {negative}{positive} point at infinity of φ be u and suppose $u \notin \Lambda(\Omega)$. There exists a neighborhood V of u in $\mathcal{S}$ such that $V \cap \bigcup_{g \in G} Ag = \emptyset$. There exists $r_0 \in \mathfrak{R}$ such that $r \in \mathfrak{R}$ with $\{r < r_0\}\{r > r_0\}$ implies $\varphi(r) \in V$. The conclusion follows.

13.30. THEOREM. *Let Ω be a limit-partial, mobile subgroup of Σ and let $G = \Omega \mid M$. Let $X^* = \bigcup_{\varphi \in \Phi} O_\varphi$ where Φ is the set of all geodesics parameterized by arclength in $\mathfrak{M}$ such that their points at infinity belong to $\Lambda(\Omega)$. Let X_G^* be the trace of X_G on X^*. Let $\mathcal{G}_G^*$ be the restriction to X_G^* of $\mathcal{G}_G$. Then:*

(1) *The set of all $\mathcal{G}_G^*$-periodic points of X_G^* is dense in X_G^*.*

(2) *$\mathcal{G}_G^*$ is transitive.*

PROOF. Use 13.24.

13.31. HOROCYCLE PARTITION FLOWS. Let Ω be a subgroup of Σ^+ and let $G = \Omega \mid M$. We observe that dG is a uniformly equicontinuous homeomorphism group of X and the horocycle flow $\mathcal{H}$ is dG-orbit preserving. Define the partition $X_G = [x\, dG \mid x \in X]$ of X and let X_G be provided with its partition uniformity. Let $\mathcal{H}_G = (X_G, \mathfrak{R}, \kappa_G)$, called the *horocycle partition flow induced by* G, be the partition flow on X_G induced by $\mathcal{H}$.

13.32. DEFINITION. Let $r \in \mathfrak{R}$. Let $x \in X$ and let $\sigma_x \in \Sigma$ be the elliptic transformation such that $(x\mu)\sigma_x = x\mu$ and such that $d\sigma_x$ rotates x through the angle r in the positive sense. The transformation $(x\, d\sigma_x \mid x \in X)$ is an analytic homeomorphism of X onto X which we call a *rotor* and denote by ρ^r.

13.33. REMARK. Let $\eta : \mathfrak{R} \to \mathfrak{M}$ be an analytic curve parameterized by arclength in $\mathfrak{M}$ such that rng η is a horocycle. Let $f(s) = h(\eta(0), \eta(s))$ $(s \in \mathfrak{R})$. Let $s \in \mathfrak{R}$ and let φ be a geodesic parameterized by arclength in $\mathfrak{M}$ such that $\eta(0) = \varphi(0)$ and $\eta(s) = \varphi(r)$ for some $r \in \mathfrak{R}^+$. Let $\alpha(s)$ $(s \in \mathfrak{R}^+)$ be the smallest positive angle from $\dot\varphi(r)$ to the unitangent externally normal to rng η at $\eta(s)$. Let $\alpha(s) = -\alpha(s)$ $(s \in \mathfrak{R}^-)$. Then the function $\alpha : s \to \alpha(s)$ $(s \in \mathfrak{R}, s \neq 0)$ is analytic with $\lim_{|s| \to \infty} \alpha(s) = 0$, the function $f : s \to f(s)$ $(s \in \mathfrak{R})$ is analytic with $\lim_{|s| \to \infty} f(s) = +\infty$, and $s \in \mathfrak{R}$ with $s \neq 0$ implies

$$\rho^\pi \rho^{\alpha(s)} \gamma^{f(s)} \rho^{\alpha(s)} = \kappa^s \tag{7}$$

where $\{\gamma^s\}\{\kappa^s\}$ is the s-transition of the {geodesic}{horocycle} flow in X.

13.34. THEOREM. *Let Ω be a limit-entire, mobile subgroup of Σ^+ and let $G = \Omega \mid M$. Then the horocycle partition flow induced by G is transitive.*

PROOF. Let θ be the projection of X onto X_G. Let $\mathcal{A}$ and $\mathcal{B}$ be open subsets of X_G and let $A = \mathcal{A}\theta^{-1}$, $B = \mathcal{B}\theta^{-1}$. There exists an open set $C \subset A\rho^{\pi}$, an open set $D \subset B$ with D saturated with respect to dG, and $\delta \in \mathcal{R}^+$ such that $r \in \mathcal{R}$ with $|r| < \delta$ implies $C\rho^r \subset A\rho^{\pi}$ and $D\rho^r \subset B$. By 13.27, there exists an orbit under $\mathcal{G}_G$ which intersects both $C\theta$ and $D\theta$. Also by 13.27 the $\mathcal{G}_G$ periodic points of X_G are dense in X_G and thus there exists a $\mathcal{G}_G$ periodic point whose orbit intersects both $C\theta$ and $D\theta$. It follows that there exist arbitrarily large real numbers t such that $C\gamma^t \cap D \neq \emptyset$ and thus (cf. 13.33) we can choose $s \in \mathcal{R}$ such that $|\alpha(s)| < \delta$ and $C\gamma^{f(s)} \cap D \neq \emptyset$. But then $A\kappa^s = A\rho^{\pi}\rho^{\alpha(s)}\gamma^{f(s)}\rho^{\alpha(s)} \supset C\gamma^{f(s)}\rho^{\alpha(s)}$ and thus $A\kappa^s \cap B \supset C\gamma^{f(s)}\rho^{\alpha(s)} \cap D\rho^{\alpha(s)} = (C\gamma^{f(s)} \cap D)\rho^{\alpha(s)} \neq \emptyset$, from which the conclusion follows.

COROLLARY. *Let Ω be a limit-entire, mobile subgroup of Σ^+ and let $G = \Omega \mid M$. Then there exists a point of X_G which is transitive under the horocycle partition flow.*

PROOF. The proof is similar to that of Corollary 13.27.

13.35. REMARK. Let Ω be a subgroup of Σ^+ and let $G = \Omega \mid M$. Let θ be the projection of X onto X_G and let $\{\overset{s}{\gamma}\}\{\overset{s}{\kappa}\}\{\overset{s}{\gamma}_G\}\{\overset{s}{\kappa}_G\}$ be the s-transition of the {geodesic flow of $\mathfrak{M}$}{horocycle flow of $\mathfrak{M}$}{geodesic partition flow in X_G induced by G}{horocycle partition flow in X_G induced by G}. Let $x \in X$, let $x\theta = x_G$ and let $t, s \in \mathcal{R}$. Then $x\overset{s}{\gamma}\overset{te^s}{\kappa} = x\overset{t}{\kappa}\overset{s}{\gamma}$ and $x_G\overset{s}{\gamma}_G\overset{te^s}{\kappa}_G = x_G\overset{t}{\kappa}_G\overset{s}{\gamma}_G$.

If $x_G \in X_G$ is {periodic}{transitive} under the horocycle partition flow $\mathcal{H}_G$ and $r \in \mathcal{R}$, then $x_G\overset{r}{\gamma}_G$ is {periodic}{transitive} under $\mathcal{H}_G$.

13.36. LEMMA. *Let Ω be a limit-entire, mobile subgroup of Σ^+ and let $G = \Omega \mid M$. Let $\overset{t}{\kappa}_G$ be the t-transition of the horocycle partition flow in X_G induced by G. Let $\mathcal{A} \subset X_G$ be the orbit of a periodic point of the geodesic partition flow $\mathcal{G}_G$ and let $\mathcal{B} = \bigcup_{t \in \mathcal{R}} \mathcal{A}\overset{t}{\kappa}_G$. Then $\mathcal{B} = X_G$.*

PROOF. Let $x \in \mathcal{A}\theta^{-1}$. Let φ be the geodesic parameterized by arclength in M such that $x = \dot{\varphi}(0)$. The range of φ is the axis of some element of G and let u be the negative point at infinity of φ. Let B be the set of unitangents of X which are externally orthogonal to the horocycles with u as point at infinity. Then $\mathcal{B} = B\theta$. Since, by 13.24(1), $\overline{u\Omega} = U$, the set $B\,dG$ is dense in X and since $\mathcal{B} = B\theta = (B\,dG)\theta$ it follows that $\mathcal{B}$ is dense in X_G. The proof is completed.

13.37. THEOREM. *Let Ω be a limit-entire, mobile subgroup of Σ^+ and let $G = \Omega \mid M$. Let $a \in X_G$ and let a be a periodic point of the geodesic partition flow $\mathcal{G}_G$. Then a is a transitive point of the horocycle partition flow $\mathcal{H}_G$.*

PROOF. It follows from 13.34(Corollary) that we can choose $b \in X_G$ such that b is transitive under the horocycle partition flow $\mathcal{H}_G$. Then (cf. 13.35) $s \in \mathcal{R}$ implies that $b\overset{s}{\gamma}_G$ is transitive under $\mathcal{H}_G$. Let $p \in \mathcal{R}^+$ be such that $a\overset{p}{\gamma}_G = a$ and let $\mathfrak{D}$ be an open subset of X_G. There exists $r \in \mathcal{R}^+$ such that $s \in \mathcal{R}$ with

$-2p \leqq s \leqq 2p$ implies $[b\gamma_G^s\kappa_G^t \mid t \in \mathfrak{R}, |t| \leqq r] \cap \mathfrak{D} \neq \emptyset$. There exists a neighborhood $\mathfrak{F}$ of b such that $f \in \mathfrak{F}$ and $s \in \mathfrak{R}$ with $-2p \leqq s \leqq 2p$ implies $[f\gamma_G^s\kappa_G^t \mid t \in \mathfrak{R}, |t| \leqq r] \cap \mathfrak{D} \neq \emptyset$. According to 13.36 there exists $r_1 \in \mathfrak{R}$ with $0 \leqq r_1 \leqq p$ and $r_2 \in \mathfrak{R}$ such that $a\gamma_G^{r_1}\kappa_G^{r_2} \in \mathfrak{F}$. Thus there exists $r_3 \in \mathfrak{R}$ such that $a\gamma_G^{r_1}\kappa_G^{r_2}\gamma_G^{-r_1}\kappa_G^{r_3} \in \mathfrak{D}$ and consequently (cf. 13.07(6))

$$a\kappa_G^{r_2e^{-r_1}+r_3} \in \mathfrak{D}.$$

The conclusion of the theorem follows.

13.38. LEMMA. *Let Ω be a limit-entire, mobile subgroup of Σ and let $G = \Omega \mid M$. Let $\Omega^+ = \Omega \cap \Sigma^+$ and let $G^+ = \Omega^+ \mid M$. Then Ω^+ is a limit-entire, mobile subgroup of Σ and the geodesic partition flow $\mathcal{G}_G$ induced by G is regionally mixing if the geodesic partition flow $\mathcal{G}_{G^+}^+$ induced by G^+ is regionally mixing.*

PROOF. It is clear that Ω^+ is a subgroup of Σ. If $\sigma \in \Omega$, then $\sigma^2 \in \Omega^+$. It follows from 13.24 that if A and B are open arcs of U, there exists $\omega \in \Omega$ such that ω has axis L with endpoints a, b such that $a \in A$ and $b \in B$. But then $\omega^2 \in \Omega^+$ and ω^2 has L as axis. We infer that Ω^+ is limit-entire and mobile.

Let θ be the projection of X onto X_G, let $\mathfrak{A}$, $\mathfrak{B}$ be open sets in X_G and let $A = \mathfrak{A}\theta^{-1}$, $B = \mathfrak{B}\theta^{-1}$. Let θ^+ be the projection of X onto X_{G^+}, let $\mathfrak{A}^+ = (\mathfrak{A}\theta^{-1})\theta^+$ and let $\mathfrak{B}^+ = (\mathfrak{B}\theta^{-1})\theta^+$. We assume that the geodesic partition flow $\mathcal{G}_{G^+}^+$ induced by G^+ is regionally mixing and thus there exists $s \in \mathfrak{R}^+$ such that $t \in \mathfrak{R}, |t| > s$ implies $\mathfrak{A}^+\gamma_{G^+}^t \cap \mathfrak{B}^+ \neq \emptyset$, where $\gamma_{G^+}^t$ denotes the t-transition of the geodesic partition flow $\mathcal{G}_{G^+}^+$. But then $t \in \mathfrak{R}$, $|t| > s$, implies $(\mathfrak{A}\theta^{-1})\gamma^t \cap (\mathfrak{B}\theta^{-1}) \neq \emptyset$, whence $\mathfrak{A}\gamma_G^t \cap \mathfrak{B} \neq \emptyset$. The proof is completed.

13.39. THEOREM. *Let Ω be a limit-entire, mobile subgroup of Σ and let $G = \Omega \mid M$. Then the geodesic partition flow $\mathcal{G}_G$ induced by G is regionally mixing.*

PROOF. Let Ω be a limit-entire, mobile subgroup of Σ and let $G = \Omega \mid M$. In view of 13.38 we can assume that $\Omega \subset \Sigma^+$.

Let $\mathfrak{A}$ and $\mathfrak{B}$ be open subsets of X_G. It follows from 13.27(1) that we can choose $a \in \mathfrak{A}$ and $p \in \mathfrak{R}^+$ such that $a\gamma_G^p = a$. Let $r \in \mathfrak{R}$ be such that $0 \leqq r \leqq p$. From 13.37 we infer that $a\gamma_G^r$ is transitive under $\mathcal{H}_G$ and thus there exists $t(r) \in \mathfrak{R}$ such that $a\gamma_G^r\kappa_G^{t(r)} \in \mathfrak{B}$. We can choose $\delta(r) \in \mathfrak{R}$ with $\delta(r) > 0$ such that $s \in \mathfrak{R}$ with $|s - r| \leqq \delta(r)$ implies $[a\gamma_G^s\kappa_G^t \mid t \in \mathfrak{R}, |t| \leqq |t(r)|] \cap \mathfrak{B} \neq \emptyset$. A finite number of the intervals $[s \mid s \in \mathfrak{R}, |s - r| \leqq \delta(r)]$ $(r \in \mathfrak{R})$ cover the interval $[r \mid r \in \mathfrak{R}, 0 \leqq r \leqq p]$ and thus there exists $t_a \in \mathfrak{R}^+$ such that $s \in \mathfrak{R}$ implies $[a\gamma_G^s\kappa_G^t \mid t \in \mathfrak{R}, |t| \leqq t_a] \cap \mathfrak{B} \neq \emptyset$.

Since $\mathfrak{A}$ is open, there exists $\delta \in R^+$ such that $[a\kappa_G^t \mid t \in \mathfrak{R}, |t| \leqq \delta] \subset \mathfrak{A}$. Let $s_0 \in \mathfrak{R}^+$ be such that $\delta e^{s_0} > t_a$ and let $s \in \mathfrak{R}$ with $s > s_0$. Then

$$\mathfrak{B} \cap \mathfrak{A}\gamma_G^s \supset \mathfrak{B} \cap ([a\kappa_G^t \mid t \in \mathfrak{R}, |t| \leqq \delta]\gamma_G^s) =$$

$$\mathfrak{B} \cap [a\gamma_G^s\kappa_G^{te^s} \mid t \in \mathfrak{R}, |t| \leqq \delta] \supset$$

$$\mathfrak{B} \cap [a\gamma_G^s\kappa_G^t \mid t \in \mathfrak{R}, |t| \leqq t_a] \neq \emptyset.$$

It remains to prove that there exists $s_1 \in \mathcal{R}$ such that $t < -s_1$, $t \in \mathcal{R}$, implies $\mathcal{A}\gamma_G^t \cap \mathcal{B} \neq \emptyset$. Let θ be the projection of X onto X_G and let $A = \mathcal{A}\theta^{-1}$, $B = \mathcal{B}\theta^{-1}$. The sets $A\rho^{\pi}$ and $B\rho^{\pi}$ are open and it follows from the first part of the proof that there exists $s_1 \in \mathcal{R}^+$ such that $s > s_1$ implies $(A\rho^{\pi})\gamma^s \cap B\rho^{\pi} \neq \emptyset$. But $s \in \mathcal{R}$ and $x \in X_G$ implies $x\rho^{\pi}\gamma^s\rho^{\pi} = x\gamma^{-s}$. Thus $s > s_1$ implies $A\gamma^{-s} \cap B = (A\rho^{\pi}\gamma^s\rho^{\pi}) \cap B = (A\rho^{\pi}\gamma^s \cap B\rho^{\pi})\rho^{\pi} \neq \emptyset$, or, equivalently, $t \in \mathcal{R}$, $t < -s_1$, implies $A\gamma^t \cap B \neq \emptyset$, and thus $\mathcal{A}\gamma_G^t \cap \mathcal{B} \neq \emptyset$. The proof is completed.

13.40. THEOREM. *Let Ω be a limit-entire, mobile subgroup of Σ^+ and let $G = \Omega \mid M$. Then the horocycle partition flow induced by G is regionally mixing.*

PROOF. Let θ be the projection of X onto X_G. Let $\mathcal{A}$ and $\mathcal{B}$ be open subsets of X_G and let $A = \mathcal{A}\theta^{-1}$, $B = \mathcal{B}\theta^{-1}$. There exists an open set $C \subset A\rho^{\pi}$, an open set $D \subset B$ with D saturated with respect to dG, and $\delta \in \mathcal{R}$ with $\delta > 0$, such that $r \in \mathcal{R}$ with $|r| < \delta$ implies $C\rho^r \subset A\rho^{\pi}$ and $D\rho^r \subset B$. By 13.39 the geodesic partition flow $\mathcal{G}_G$ is regionally mixing and we can choose $s_0 \in \mathcal{R}$ such that $|s| > s_0$ implies $|\alpha(s)| < \delta$ and $C\gamma^{f(s)} \cap D \neq 0$ (cf. 13.33). But then $|s| > s_0$ implies $A\kappa^s \cap B \supset A\rho^{\pi}\rho^{\alpha(s)}\gamma^{f(s)}\rho^{\alpha(s)} \cap D\rho^{\alpha(s)} \supset (C\gamma^{f(s)} \cap D)\rho^{\alpha(s)} \neq \emptyset$, from which the conclusion of the theorem follows.

13.41. COMPLETE TWO-DIMENSIONAL RIEMANNIAN MANIFOLDS OF CONSTANT NEGATIVE CURVATURE. Let $\mathfrak{N}$ be a complete two-dimensional analytic Riemannian manifold of constant negative curvature -1. There exists (cf. H. Hopf [1]) a group G of isometries of $\mathfrak{M}$ onto $\mathfrak{M}$ such that $\mathfrak{M}$ is the universal covering manifold of $\mathfrak{N}$ with G the covering group. The two-dimensional Riemannian manifold $\mathfrak{M}_G$ obtained by partitioning $\mathfrak{M}$ by G is isometric to $\mathfrak{N}$.

The group G has the property that $p \in \mathfrak{M}$ implies the existence of a neighborhood W of p such that $pg \in W$, $g \in G$, only if g is the identity mapping. A group of isometries of $\mathfrak{M}$ with this property will be said to be *discrete in* $\mathfrak{M}$.

Let F be a group of isometries of $\mathfrak{M}$ onto $\mathfrak{M}$ which is discrete in $\mathfrak{M}$. The two-dimensional Riemannian manifold obtained by partitioning $\mathfrak{M}$ by F is then a complete two-dimensional analytic Riemannian manifold of constant negative curvature -1. Thus the problem of constructing the class of such manifolds is equivalent to the problem of constructing the class of groups of isometries of $\mathfrak{M}$ which are discrete in $\mathfrak{M}$.

13.42. LEMMA. *Let Ω be a subgroup of Σ with* crd $\Lambda(\Omega) > 2$ *and let $G = \Omega \mid M$ be discrete in $\mathfrak{M}$. Then Ω is mobile.*

PROOF. We assume that Ω is not mobile and thus that there exists $u \in U$ such that $u\Omega = u$. Let $\omega_0 \in \Omega$ be the identity mapping. Since G is discrete in $\mathfrak{M}$, $\omega \in \Omega$, $\omega \neq \omega_0$, implies that ω is either parabolic, hyperbolic or a paddle motion. If all $\omega \in \Omega$ other than ω_0 were parabolic, the set $\Lambda(\Omega)$ would consist of $[u]$, contrary to hypothesis. Thus there exists $\omega_1 \in \Omega$ with axis L_1 and L_1 must have u as one of its endpoints. If L_1 were the axis of every member of Ω other than ω_0, it would follow that $\Lambda(\Omega) = 2$, which is not the case. Let $\omega \in \Omega$ such that L_1

is not the axis of ω. If ω does not have an axis, then ω is parabolic and $L_1\omega$ is an axis of $\omega^{-1}\omega_1\omega \in \Omega$ with $L_1\omega \neq L_1$. Thus there exists $\omega_2 \in \Omega$ with axis L_2 such that $L_1 \neq L_2$. Since $u\Omega = u$, L_1 and L_2 must have u as common endpoint.

Let $p_1 \in L_1$, let $p_2 \in L_2$ and let $\delta \in \mathcal{R}^+$. There exists $m \in \mathcal{I}$ such that $0 < D(p_1\omega_1^m, L_2) < \delta$ and thus there exists $n \in \mathcal{I}$ such that the h-distance from $p_1\omega_1^m\omega_2^n$ to some point of the h-line segment joining p_2 and $p_2\omega_2$ is positive and less than δ. Since δ can be chosen arbitrarily small there exists a sequence $\{\omega_k^* \mid k \in \mathcal{I}^+, \omega_k^* \in \Omega\}$ such that $p_1\omega_i^* \neq p_1\omega_j^*$, $i \neq j$, and such that the sequence $\{p_1\omega_k^* \mid k \in \mathcal{I}^+\}$ converges to a point $p \in L_2$. But this implies that G is not discrete in $\mathfrak{M}$. The proof of the lemma is completed.

13.43. Theorem. *Let Ω be a subgroup of Σ, let $G = \Omega \mid M$ and let G be discrete in $\mathfrak{M}$. Let ω_0 be the identity mapping of G. Then exactly one of the following statements is valid.*

(1) crd $\Lambda(\Omega) = 0$; $\Omega = [\omega_0]$.

(2) crd $\Lambda(\Omega) = 1$; *there exists $\omega \in \Omega$ such that ω is parabolic with fixed point $\Lambda(\Omega)$ and $\Omega = \{\omega^n \mid n \in \mathcal{I}\}$.*

(3) crd $\Lambda(\Omega) = 2$; *there exists $\omega \in \Omega$ such that ω is hyperbolic with fixed points $\Lambda(\Omega)$ and $\Omega = \{\omega^n \mid n \in \mathcal{I}\}$.*

(4) crd $\Lambda(\Omega) = 2$; *there exists $\omega \in \Omega$ such that ω is a paddle motion with fixed points $\Lambda(\Omega)$ and $\Omega = \{\omega^n \mid n \in \mathcal{I}\}$.*

(5) crd $\Lambda(\Omega) > 2$; *Ω is mobile and limit-partial.*

(6) crd $\Lambda(\Omega) > 2$; *Ω is mobile and limit-entire.*

Proof. Since G is discrete in $\mathfrak{M}$, $\omega \in \Omega$ with $\omega \neq \omega_0$ implies that ω is parabolic, hyperbolic or a paddle motion. Thus if $\Lambda(\Omega) = \emptyset$, it follows that $\Omega = [\omega_0]$.

Suppose crd $\Lambda(\Omega) = 1$. Let $u = \Lambda(\Omega)$. Then $\omega \in \Omega$ with $\omega \neq \omega_0$ implies that ω is parabolic with fixed point u. Let $p \in \mathfrak{M}$ and let H be the horocycle with point at infinity u and such that $p \in H$. Then $p\Omega \subset H$ and since G is discrete in $\mathfrak{M}$, there exists $p_1 \in p\Omega$ such that $p_1 \neq p$ and $h(p_1, p) \leqq h(q, p)$ for all $q \in p\Omega$, $q \neq p$. Let $p_1 = p\omega$ with $\omega \in \Omega$. Then ω is parabolic with fixed point u and $\Omega = \{\omega^n \mid n \in \mathcal{I}\}$.

Suppose crd $\Lambda(\Omega) = 2$. Let $\Lambda(\Omega) = [u, v]$. Since $\Lambda(\Omega)$ is invariant under Ω and G is discrete in $\mathfrak{M}$, $\omega \in \Omega$ implies that $u\omega = u$ and $v\omega = v$. Let L be the h-line with points at infinity u and v and let $p \in L$. Since G is discrete in $\mathfrak{M}$, there exists $p_1 \in p\Omega$, $p_1 \neq p$, $p_1 \in L$ such that $h(p_1, p) \leqq h(q, p)$ for all $q \in p\Omega$, $q \neq p$. Let $p_1 = p\omega$ with $\omega \in \Omega$. Then $\Omega = \{\omega^n \mid n \in \mathcal{I}\}$ and ω is either a hyperbolic transformation or a paddle motion with u and v as fixed points in either case.

If crd $\Lambda(\Omega) > 2$, it follows from 13.42 that Ω is mobile.

It follows from 13.15 that all possible cases have been considered. The proof is completed.

13.44. Remark. Let Ω be a subgroup of Σ, let $G = \Omega \mid M$ and let G be discrete in $\mathfrak{M}$. Then G is countable and if $p \in M$, $\omega_1, \omega_2 \in \Omega$, $p\omega_1 = p\omega_2$ then $\omega_1 = \omega_2$.

13.45. DEFINITION. Let Ω be a subgroup of Σ, let $G = \Omega \mid M$ and let G be discrete in $\mathfrak{M}$. Let p_0 be the origin O and let $p_0G = [p_i \mid i = 0, 1, 2, \cdots]$, where $p_i \neq p_j$ provided $i \neq j$. For $n \in \mathscr{I}^+$ define $R_n = [p \mid p \in \mathfrak{M}, h(p, p_0) < h(p, p_n)]$. The set $R = \bigcap_{n=1}^{\infty} R_n$ is the *fundamental region of* G.

13.46. REMARK. Let Ω be a subgroup of Σ, let $G = \Omega \mid M$, let G be discrete in $\mathfrak{M}$ and let R be the fundamental region of G. Then:

(1) R is a non-vacuous open subset of $\mathfrak{M}$.

(2) R is h-convex in the sense that $p, q \in R$ implies that the h-line segment joining p and q lies in R.

(3) If $g_1, g_2 \in G$ and $Rg_1 \cap Rg_2 \neq \emptyset$ then $g_1 = g_2$.

(4) Corresponding to any compact subset A of $\mathfrak{M}$ there exists a finite subset E of G such that $A \subset \bigcup_{g \in E} \overline{Rg}$.

(5) $\mathcal{A}_h(R) = \mathcal{A}_h(\overline{R})$.

13.47. THEOREM. *Let Ω be a subgroup of Σ, let $G = \Omega \mid M$, let G be discrete in $\mathfrak{M}$ and let R be the fundamental region of G. If R is of finite h-area, then* crd $\Lambda(\Omega) > 2$ *and Ω is mobile and limit-entire.*

PROOF. It follows from 13.43 that if crd $\Lambda(\Omega) \leqq 2$ then R is not of finite h-area. Thus we can assume that crd $\Lambda(\Omega) > 2$ and by 13.42 Ω is mobile. We show that Ω is limit-entire.

Let $\delta \in \mathcal{R}^+$. There exists $r \in \mathcal{R}$, $0 < r < 1$, such that if C_r denotes the circle with center O and euclidean radius r, then any h-convex subset of $\mathfrak{M}$ which is exterior to C_r has euclidean diameter less than δ. There exists a finite subset E of G such that C_r and its interior are contained in $\bigcup_{g \in E} \overline{Rg}$. Let $u \in U$ and let W be a euclidean neighborhood of u of diameter less than δ and exterior to C_r. Then $W \cap \mathfrak{M}$ is of infinite h-area and there must exist $g^* \in G$, $g^* \notin E$, such that $Rg^* \cap W \neq \emptyset$. But then Rg^* is exterior to C_r, the euclidean diameter of Rg^* is less than δ and Og^* is within euclidean distance 2δ of u. It follows that $u \in \Lambda(\Omega)$ and thus $U \subset \Lambda(\Omega)$ and Ω is limit-entire.

13.48. GEODESIC FLOWS OF TWO-DIMENSIONAL MANIFOLDS OF CONSTANT NEGATIVE CURVATURE. Let $\mathfrak{N}$ be a complete two-dimensional analytic Riemannian manifold of constant curvature -1. Let $p \in \mathfrak{N}$. *A unitangent on $\mathfrak{N}$ at p* is a unit contravariant vector at p. The *unitangent space on $\mathfrak{N}$ at p*, denoted $\mathfrak{I}(\mathfrak{N}, p)$ is the set of all unitangents on $\mathfrak{N}$ at p.

The *unitangent space on* $\mathfrak{N}$, denoted X, is $\bigcup_{p \in \mathfrak{N}} \mathfrak{I}(\mathfrak{N}, p)$. Let $x \in X$ and let x be a unitangent at $p \in \mathfrak{N}$. Let $r \in \mathcal{R}^+$ and let $A_r(x)$ be the set of all unitangents on $\mathfrak{N}$ at p and forming an angle less than r with x. Let $U_r(x)$ be the set of all unitangents obtained from A_r by parallel transport along all geodesic segments of length less than or equal to r and with initial point p. For $r \in \mathcal{R}^+$ define $\alpha_r = [(x_1, x_2) \mid x_1, x_2 \in X, x_2 \in U_r(x_1)]$. Define $\mathcal{V} = [\alpha_r \mid r \in \mathcal{R}^+]$. It is readily verified that $\mathcal{V}$ is a uniformity base. Let $\mathcal{U}$ be the uniformity generated by $\mathcal{V}$. We provide X with this uniformity and assign to X the topology induced by $\mathcal{U}$.

Let γ be the transformation of $X \times \mathcal{R}$ onto X defined as follows. Let $x \in X$

and let $s \in \mathcal{R}$. Let φ be a geodesic parameterized by arclength in $\mathfrak{N}$ such that $x = \dot{\varphi}(s_0)$ is the tangent vector to φ at $\varphi(s_0)$. Let $y = \dot{\varphi}(s + s_0)$. We define $(x, s)\gamma = y$. Then $G = (X, \mathcal{R}, \gamma)$ is a transformation group on X which is called the *geodesic flow of* $\mathfrak{N}$.

13.49. THEOREM. *Let $\mathfrak{N}$ be a complete two-dimensional analytic Riemannian manifold of constant curvature -1 and of finite area. Then the geodesic flow of $\mathfrak{N}$ is regionally transitive, regionally mixing, and the periodic orbits of the geodesic flow of $\mathfrak{N}$ are dense in the space of unitangents on $\mathfrak{N}$.*

PROOF. Use 13.27, 13.39 and 13.47.

13.50. CONSTRUCTION OF TWO-DIMENSIONAL MANIFOLDS OF CONSTANT NEGATIVE CURVATURE. As indicated in 13.41, the problem of construction of two-dimensional Riemannian manifolds of constant negative curvature -1 is equivalent to the problem of construction of groups of isometries of $\mathfrak{M}$ which are discrete in $\mathfrak{M}$. The problem can be completely solved by geometric methods involving the construction of fundamental regions (cf. Fricke-Klein [1], Koebe [1] and Löbell [1]). These manifolds include compact orientable manifolds of genus at least 2 and compact non-orientable manifolds of every topological type other than the projective plane and Klein bottle.

13.51. n-DIMENSIONAL MANIFOLDS OF CONSTANT NEGATIVE CURVATURE, $n > 2$. A large number of the results of this section can be extended to manifolds of constant negative curvature of dimension exceeding 2 and the proofs have been so designed that these extensions obtain with scarcely any modifications of the proofs given for the case of dimension 2. In particular, the extensions of the results concerning the density of the periodic geodesics and regional transitivity of the geodesic flow are valid. The extension of the concept of the horocycle flow is not immediately obvious, but mixing properties can be attained for higher dimensional manifolds (cf. E. Hopf [3]).

The construction and classification of manifolds of constant negative curvature and dimension exceeding 2 is largely an unsolved problem. Compact manifolds of constant negative curvature and of dimension 3 have been constructed by Löbell [2] and Salenius [1], while non-compact manifolds of finite volume are known to exist, but these examples appear to represent only a small number of the possibilities.

13.52. NOTES AND REFERENCES.

(13.01) This model of the hyperbolic plane is commonly associated with Poincaré due to his extensive use of it in the development of the theory of automorphic functions, although it appears to have been known earlier to Beltrami (cf. Beltrami [1]).

(13.02) Cf. Bianchi [1], p. 584.

(13.11) Cf. E. Hopf [3], p. 268.

(13.13) It is usually assumed in the definition and analysis of limit sets that $G = \Omega \mid M$ is *properly discontinuous* (cf. L. R. Ford [1]).

(13.27, 13.34, 13.39) Cf. Hedlund [2] for references to the literature.

(13.41) The expression *discrete in* $\mathfrak{M}$ replaces the more commonly used expression *properly discontinuous*. The phrase *discrete and without fixed points in* $\mathfrak{M}$ would, perhaps, be more appropriate. A group of isometries of $\mathfrak{M}$ can be topologized in various ways. It can be considered as a set of mappings of $\mathfrak{M}$ onto $\mathfrak{M}$ and assigned the compact-open topology. It can be considered as a set of mappings of X onto X and assigned the compact-open topology. It can be considered as the restriction of a subgroup of Σ and thus be topologized by defining a base $\mathcal{B}$ for the neighborhoods of the identity of Σ as follows: let $\epsilon \in \mathcal{R}^+$, let

$$B(\epsilon) = \left\{ \frac{az + \bar{c}}{cz + \bar{a}} \;\middle|\; a, c \in \mathcal{S};\, a\bar{a} - c\bar{c} = 1;\, |\, a - 1 \,| < \epsilon,\, |\, c \,| < \epsilon \right\}$$

and let $\mathcal{B} = \bigcup_{\epsilon \in R^+} B(\epsilon)$. Then G is discrete in $\mathfrak{M}$, as defined in 13.41, if and only if no element of G other than the identity has a fixed point in $\mathfrak{M}$, and G is discrete in each of the stated topologies.

14. CYLINDER FLOWS AND A PLANAR FLOW

14.01. STANDING NOTATION. Throughout this section Y denotes a topological space, $\mathcal{R}$ denotes the set of real numbers with the natural topology, $F(Y)$ denotes the set of all continuous functions on Y to $\mathcal{R}$ and $H(Y)$ denotes the set of all homeomorphisms of Y onto Y.

14.02. DEFINITION. Let $X = Y \times \mathcal{R}$ and let ν denote the *projection of X onto $\mathcal{R}$* defined by $(y, r)\nu = r$ $(y \in Y, r \in \mathcal{R})$. The subset A of X is {*bounded above*}{*bounded below*} provided $A\nu$ is {bounded above}{bounded below}, and A is *bounded* provided A is bounded both above and below.

14.03. DEFINITION. Let $\theta \in H(Y)$ and let $f \in F(Y)$. Let $X = Y \times \mathcal{R}$ and let φ be the homeomorphism of X onto X defined by $(y, r)\varphi = (y\theta, r + f(y))$ $(y \in Y, r \in \mathcal{R})$. The homeomorphism φ will be denoted $\varphi(Y, f, \theta)$ and called the *cylinder homeomorphism determined by Y, f and θ.*

14.04. DEFINITION. Let $X = Y \times \mathcal{R}$, let $s \in \mathcal{R}$ and let $\psi_s : X \to X$ be defined by $(y, r)\psi_s = (y, r + s)$ $(y \in Y, r \in \mathcal{R})$. The homeomorphism ψ_s of X onto X will be called the *translation of X by s.*

14.05. REMARK. Let $X = Y \times \mathcal{R}$, let $s \in \mathcal{R}$ and let ψ_s be the translation of X by s. Let $f \in F(Y)$, let $\theta \in H(Y)$ and let $\varphi = \varphi(Y, F, \theta)$. Let $x \in X$, let $\Phi(x)$ be the orbit of x under φ and let $\Gamma(x) = \overline{\Phi(x)}$. Then

(1) $x\varphi\psi_s = x\psi_s\varphi$.
(2) $\Phi(x)\psi_s = \Phi(x\psi_s)$.
(3) $\Gamma(x)\psi_s = \Gamma(x\psi_s)$.

14.06. DEFINITION. Let $f \in F(Y)$, let $\theta \in H(Y)$ and let $\varphi = \varphi(Y, f, \theta)$. Let $\{A^-\}\{A^+\}\{\Omega^-\}\{\Omega^+\}$ denote the set of all $x \in X = Y \times \mathcal{R}$ such that {the negative semiorbit of x under φ is bounded below}{the negative semiorbit of x under φ is bounded above}{the positive semiorbit of x under φ is bounded below}{the positive semiorbit of x under φ is bounded above}. Let $\{\alpha^-\}\{\alpha^+\}$ $\{\omega^-\}\{\omega^+\}$ denote the set of all $x \in X$ such that

$$\{\lim_{n\to-\infty} x\varphi^n\nu = -\infty\}\{\lim_{n\to-\infty} x\varphi^n\nu = +\infty\}\{\lim_{n\to+\infty} x\varphi^n\nu = -\infty\}\{\lim_{n\to+\infty} x\varphi^n\nu = +\infty\}.$$

Let $B^+ = A^+ \cap \Omega^+$. Let $B^- = A^- \cap \Omega$. The collection of sets $[A^-, A^+, \Omega^-, \Omega^+, \alpha^-, \alpha^+, \omega^-, \omega^+, A^{-\prime}, A^{+\prime}, \Omega^{-\prime}, \Omega^{+\prime}, B^+, B^-]$ will be denoted by $D(\varphi)$.

14.07. REMARK. Let $\theta \in H(Y)$, let Y be a minimal orbit-closure under θ, let $f \in F(Y)$ and let $\varphi = \varphi(Y, f, \theta)$. Let $s \in \mathcal{R}$ and let ψ_s be the translation of X by s. Let $D \in D(\varphi)$. Then D is invariant under φ, D is invariant under ψ_s and if $D \neq \emptyset$ then D is dense in X.

14.08. LEMMA. *Let Y be a compact metric space, let $\theta \in H(Y)$, let $f \in F(Y)$, let $\varphi = \varphi(Y, f, \theta)$ and let ν be the projection of $X = Y \times \mathcal{R}$ onto $\mathcal{R}$. For each $i \in \mathcal{I}^+$ let there exist $x_i \in X$ and $n_i \in \mathcal{I}^+$ such that $\{x_i\varphi^{n_i}\nu - x_i\nu > i\}\{x_i\varphi^{n_i}\nu - x_i\nu < -i\}$. Then*

(1) $\{A^+ \neq \emptyset\}\ \{A^- \neq \emptyset\}$.

(2) $\{\Omega^- \neq \emptyset\}\ \{\Omega^+ \neq \emptyset\}$.

PROOF. It is sufficient to prove the first reading.

We assume that corresponding to $i \in \mathcal{I}^+$ there exists $x_i \in X$ and $n_i \in \mathcal{I}^+$ such that $x_i\varphi^{n_i}\nu - x_i\nu > i$. Corresponding to $i \in \mathcal{I}^+$ let $p_i, q_i \in \mathcal{I}^+$ be so chosen that $0 \leqq p_i < q_i \leqq n_i$ and $x_i\varphi^{q_i}\nu - x_i\varphi^{p_i}\nu = \sup_{0 \leqq m < n \leqq n_i} (x_i\varphi^n\nu - x_i\varphi^m\nu)$. Then $x_i\varphi^{q_i}\nu - x_i\varphi^{p_i}\nu > i$ and $k \in \mathcal{I}$, $p_i \leqq k \leqq q_i$ implies $x_i\varphi^{p_i}\nu \leqq x_i\varphi^k\nu \leqq x_i\varphi^{q_i}\nu$. Since Y is compact and f is continuous on Y there exists $L \in \mathcal{R}^+$ such that $y \in Y$ implies $|\,yf\,| \leqq L$ and thus $x \in X$ implies $|\,x\varphi\nu - x\nu\,| \leqq L$. It follows that $i \in \mathcal{I}^+$ implies $x_i\varphi^{q_i}\nu - x_i\varphi^{p_i}\nu \leqq (q_i - p_i)L$ and thus $q_i - p_i > i/L$, whence $\lim_{i \to +\infty} (q_i - p_i) = +\infty$.

Let $s_i = x_i\varphi^{q_i}\nu$ and let $x_i^* = x_i\varphi^{q_i}\psi_{-s_i}$, $i \in \mathcal{I}^+$. Then $x_i^*\nu = 0$ $(i \in \mathcal{I}^+)$ and $x_i^*\varphi^k\nu = x_i\varphi^{q_i}\psi_{-s_i}\varphi^k\nu = x_i\varphi^{q_i+k}\nu - s_i = x_i\varphi^{q_i+k}\nu - x_i\varphi^{q_i}\nu \leqq 0$, $k = 0, -1, \cdots, p_i - q_i$. Since Y is compact, the sequence $x_1^*, x_2^*, \cdots$ contains a subsequence converging to a point $x^* \in X$ and $n \in \mathcal{I}^-$ implies $x^*\varphi^n\nu \leqq 0$. Thus $x^* \in A^+$ and (1) is proved.

Let $t_i = x_i\varphi^{p_i}\nu$ and let $x_i' = x_i\varphi^{p_i}\psi_{-t_i}$, $i \in \mathcal{I}^+$. Then $x_i'\nu = 0$ $(i \in \mathcal{I}^+)$ and $x_i'\varphi^k\nu = x_i\varphi^{p_i}\psi_{-t_i}\varphi^k\nu = x_i\varphi^{p_i+k}\nu - t_i = x_i\varphi^{p_i+k}\nu - x_i\varphi^{p_i}\nu \geqq 0$, $k = 0, 1, \cdots, q_i - p_i$. The sequence $x_1', x_2', \cdots$ contains a subsequence converging to a point $x' \in X$ and $n \in \mathcal{I}^+$ implies $x'\varphi^n\nu \geqq 0$. Thus $x' \in \Omega^-$ and (2) is proved.

14.09. LEMMA. *Let Y be a compact metric space, let $\theta \in H(Y)$, let $f \in F(Y)$, let $\varphi = \varphi(Y, f, \theta)$ and let ν be the projection of $X = Y \times \mathcal{R}$ onto $\mathcal{R}$. For each $i \in \mathcal{I}^+$ let there exist $x_i \in X$ and $m_i, n_i \in \mathcal{I}^+$, with $0 < m_i < n_i$, such that $\{x_i\varphi^{m_i}\nu - x_i\nu > i\}\{x_i\varphi^{m_i}\nu - x_i\nu < -i\}$ and $\{x_i\varphi^{m_i}\nu - x_i\varphi^{n_i}\nu > i\}\{x_i\varphi^{m_i}\nu - x_i\varphi^{n_i}\nu < -i\}$. Then $\{B^+ \neq \emptyset\}\{B^- \neq \emptyset\}$.*

PROOF. It is sufficient to prove the first reading.

We assume that for each $i \in \mathcal{I}^+$ there exists $x_i \in X$ and $m_i, n_i \in \mathcal{I}^+$ such that $0 < m_i < n_i$ and $x_i\varphi^{m_i}\nu - x_i\nu > i$, $x_i\varphi^{m_i}\nu - x_i\varphi^{n_i}\nu > i$. Let $p_i \in \mathcal{I}^+$ be such that $0 \leqq p_i \leqq n_i$ and $x_i\varphi^{p_i}\nu = \sup_{0 \leqq n \leqq n_i} x_i\varphi^{n_i}$. Then $x_i\varphi^{p_i}\nu - x_i\nu > i$ and $x_i\varphi^{p_i}\nu - x_i\varphi^{n_i}\nu > i$, $i \in \mathcal{I}^+$, in consequence of which $\lim_{i \to +\infty} p_i = +\infty$ and $\lim_{i \to +\infty} (n_i - p_i) = +\infty$. Let $s_i = x_i\varphi^{p_i}\nu$, let ψ_{-s_i} be the translation of X by $-s_i$ and let $x_i^* = x_i\varphi^{p_i}\psi_{-s_i}$, $i \in \mathcal{I}^+$. Then $x_i^*\nu = 0$ and $x_i^*\varphi^k\nu = x_i\varphi^{p_i}\psi_{-s_i}\varphi^k\nu = x_i\varphi^{p_i+k}\nu - x_i\varphi^{p_i}\nu$, $i \in \mathcal{I}^+$, $k \in \mathcal{I}$, and consequently $x_i^*\varphi^k\nu \leqq 0$, $k = -p_i, -p_i + 1, \cdots, n_i - p_i$. Since Y is compact, the sequence $x_1^*, x_2^*, \cdots$ contains a subsequence converging to a point $x^* \in X$ and $n \in \mathcal{I}$ implies $x^*\varphi^n\nu \leqq 0$. Thus $x^* \in B^+$, and the proof is completed.

14.10. LEMMA. *Let $\theta \in H(Y)$, let $f \in F(Y)$, let $X = Y \times \mathcal{R}$ and let $\varphi = \varphi(Y, f, \theta)$. Let $x = (y, r) \in X$, let $\Gamma(x)$ be the orbit-closure of x under φ and* *let there exist* $s \in \mathcal{R}$, $s \neq 0$, *such that* $x\psi_s = (y, r + s) \in \Gamma(x)$. *Then* $[x\psi_s^n \mid n \in \mathcal{J}^+] = [(y, r + ns) \mid n \in \mathcal{J}^+] \subset \Gamma(x)$.

PROOF. Let $x\psi_s \in \Gamma(x)$ and let $n \in \mathcal{J}^+$. Since $\Gamma(x)$ is closed and invariant under φ, it follows that $\Gamma(x\psi_s) \subset \Gamma(x)$. With the aid of 14.05(3), it follows by induction that $\Gamma(x\psi_s^n) \subset \Gamma(x)$, whence $x\psi_s^n \in \Gamma(x)$ and $[x\psi_s^n \mid n \in \mathcal{J}^+] \subset \Gamma(x)$.

14.11. THEOREM. *Let* $\theta \in H(Y)$, *let* Y *be a compact minimal orbit-closure under* θ, *let* $f \in F(Y)$ *and let* $\varphi = \varphi(Y, f, \theta)$. *Then the following statements are equivalent*:

(1) *There exists* $x \in X$ *such that at least one of the semiorbits of* x *under* φ *is bounded.*

(2) $x \in X$ *implies the existence of* $g_x \in F(Y)$ *such that the orbit-closure of* x *under* φ *is* g_x.

(3) *There exists* $g \in F(Y)$ *such that* $f(y) = g(y\theta) - g(y)$ $(y \in Y)$.

(4) $\sum_{p=0}^{n-1} f(y\theta^p)$ *is bounded on* $Y \times \mathcal{J}^+$.

(5) φ *is pointwise almost periodic.*

PROOF. Assume (1). We prove (2). Let $x_0 \in X$ and let the positive semiorbit of x_0 under φ be bounded. That is, there exist $a, b \in \mathcal{R}$ such that if ν denotes the projection of X onto $\mathcal{R}$ and $n \in \mathcal{J}^+$, then $a \leqq x_0\varphi^n\nu \leqq b$. Let $\omega(x_0)$ be the ω-limiting set of x_0 under φ. Then $\omega(x_0)\nu$ is contained in the closed interval (a, b), $\omega(x_0)$ is a compact invariant nonvacuous set and $\omega(x_0)$ contains a minimal set M.

Let $x^* = (y^*, r^*) \in M$. Since M is invariant under φ, we have $[x^*\varphi^n \mid n \in \mathcal{J}^+] = [(y^*\theta^n, r^* + \sum_{p=0}^{n-1} f(y^*\theta^p)) \mid n \in \mathcal{J}^*] \subset M$ and since $M\nu \subset (a, b)$, it follows that $n \in \mathcal{J}^+$ implies $a \leqq r^* + \sum_{p=0}^{n-1} f(y^*\theta^p) \leqq b$. Since Y is a compact minimal orbit-closure under θ, it follows that $[y^*\theta^n \mid n \in \mathcal{J}^+]$ is dense in Y and thus, if $R(y)$ denotes the set $[(y, r) \mid r \in \mathcal{R}]$, it follows that $y \in Y$ implies $M \cap R(y) \neq \emptyset$. For each $y \in Y$, the set $M \cap R(y)$ consists of exactly one point. For if $(y, r) \in M$ and $(y, r + s) \in M$, with $s \neq 0$, it follows from 14.10 that $[(y, r + ns) \mid n \in \mathcal{J}^+] \subset M$, which is not possible, since M is bounded.

Let $g : Y \to R$ be defined by $(y, g(y)) \in M$, $y \in Y$. Then $g(y)$ is uniquely defined for each $y \in Y$, $g(y) \in \mathcal{R}$, and g is a bounded function on Y. Since $g = M = \overline{M} = \overline{g}$, it follows that g is continuous and thus $g \in F(Y)$.

Let $x = (y_0, r) \in X$. Then $(y_0, g(y_0)) \in M$, the orbit-closure of $(y_0, g(y_0))$ under φ is M, $x \in M\psi_{r-g(y_0)}$ and $M\psi_{r-g(y_0)}$ is the orbit-closure of x under φ. Let $g_x : Y \to \mathcal{R}$ be defined by $g_x(y) = r - g(y_0) + g(y)$, $y \in Y$. Then $g_x \in F(Y)$ and $g_x = M\psi_{r-g(y_0)}$, which is the orbit-closure of x under φ. Thus (2) is true for the case under consideration.

The similar proof that the existence of a bounded negative semiorbit implies (2) will be omitted.

Assume (2). We prove (3). Let $x \in X$ and let $g \in F(Y)$ be the orbit-closure of x under φ. Then $(y\theta, g(y\theta)) = (y, g(y))\varphi = (y\theta, g(y) + f(y))$, $y \in Y$, and thus $f(y) = g(y\theta) - g(y)$, $y \in Y$, which implies (3).

Assume (3). We prove (4). Let $g \in F(Y)$ exist such that $f(y) = g(y\theta) - g(y)$, $y \in Y$. Let $y \in Y$ and let $n \in \mathcal{I}^+$. Then

$$\sum_{p=0}^{n-1} f(y\theta^p) = \sum_{p=0}^{n-1} \{g(y\theta^{p+1}) - g(y\theta^p)\} = g(y\theta^n) - g(y).$$

Since Y is compact, g is bounded on Y and there exists $b \in \mathcal{R}$ such that $y \in Y$ implies $| g(y) | \leqq b$. Thus $y \in Y$ and $n \in \mathcal{I}^+$ implies

$$\left| \sum_{p=0}^{n-1} f(y\theta^p) \right| \leqq | g(y\theta^n) - g(y) | \leqq 2b,$$

which implies (4).

Clearly (4) implies that the positive semiorbit of any point $x \in X$ is bounded and thus (4) implies (1).

Assume (2). We prove (5). Let $x_0 = (y_0, r_0) \in X$ and let U be a neighborhood of x_0. Let $g_0 \in F(Y)$ be the orbit-closure of x_0 under φ. Since g_0 is continuous on Y, there exists a neighborhood V of y_0 in Y such that $y \in V$ implies $(y, g_0(y)) \in U$. Since Y is a compact minimal orbit-closure under θ, there exists a syndetic subset A of $\mathcal{I}$ such that $a \in A$ implies $y_0\theta^a \in V$. But then $a \in A$ implies that $x_0\varphi^a = (y_0\theta^a, g_0(y_0\theta^a)) \in U$ and thus x_0 is almost periodic under φ, which implies (5).

Clearly (5) implies (1).

The proof of the theorem is completed.

14.12. Remark. Let Y be a compact metric space, let $\theta \in H(Y)$ and let Y be an almost periodic minimal orbit-closure under θ. Let $y \in Y$. By 4.48 there exists a unique group structure of Y which makes Y a topological group such that (Y, π_y), where $\pi_y : \mathcal{I} \to Y$ is defined by $i\pi_y = y\varphi^i$, $i \in \mathcal{I}$, is a compactification of $\mathcal{I}$ and thus (cf. Halmos [1]) there exists a unique normalized regular Haar measure μ on Y.

14.13. Theorem. *Let Y be a compact, connected, locally connected, metric space, let $\theta \in H(Y)$, let Y be an almost periodic minimal orbit-closure under θ, let $f \in F(Y)$, let $\varphi = \varphi(Y, f, \theta)$ and let μ be the normalized Haar measure on Y. Then the following statements are equivalent:*

(1) *The discrete flow generated by φ is transitive.*

(2) *The discrete flow generated by φ is point extensively transitive.*

(3–4) *There exists $x_1 \in X$ such that that the {positive}{negative} semiorbit of x_1 is not bounded below and there exists $x_2 \in X$ such that the {positive}{negative} semiorbit of x_2 is not bounded above.*

(5–6) *There exists $x \in X$ such that neither semiorbit of x is bounded {below} {above}.*

(7) *$\int_Y f(y)\, d\mu(y) = 0$ and $\sum_{p=0}^{n-1} f(y\theta^p)$ is not bounded on $Y \times \mathcal{I}^+$.*

Proof. It follows from 9.23 that (1) and (2) are equivalent.

Clearly (2) implies (3), (4), (5) and (6).

Assume (3). We prove (1). Let $x \in \Omega^{-\prime}$. If $x \in \omega^-$, then $x \in \Omega^+$. If $x \in \Omega^{-\prime} - \omega^-$, it follows from 14.08 that $\Omega^+ \neq \emptyset$. Similarly, $\Omega^{+\prime} \neq \emptyset$ implies $\Omega^- \neq \emptyset$.

Let U and V be open subsets of $X = Y \times \mathcal{R}$. Let Y_u be an open nonvacuous subset of Y and I_u an open bounded nonvacuous interval of $\mathcal{R}$ such that $Y_u \times I_u \subset U$. Let Y_2 be an open nonvacuous subset of Y and I_v an open bounded nonvacuous interval of $\mathcal{R}$ such that $Y_2 \times I_v \subset V$. Since Y is an almost periodic minimal orbit-closure under θ, there exists an open nonvacuous set $Y_v \subset Y_2$ and a syndetic subset A of $\mathcal{I}$ such that $a \in A$ implies $Y_v\theta^a \subset Y_u$. Thus there exists $M \in \mathcal{I}^+$ such that $n \in \mathcal{I}$ implies that for some integer j, $1 \leqq j \leqq M$ and $Y_v\theta^{n+j} \subset Y_u$.

Since $\Omega^+ \neq \emptyset \neq \Omega^-$, it follows from 14.07 that each of these sets is dense in X. Let $x_+ = (y_+, r_+) \in (Y_v \times I_v) \cap \Omega^+$ and let $x_- = (y_-, r_-) \in (Y_v \times I_v) \cap \Omega^-$ such that x_+ and x_- lie in a connected subset C of $Y_v \times I_v$. We observe that $x_+ \in \Omega^{-1}$ and $x_- \in \Omega^{+1}$, for otherwise it would follow from 14.11 that $\Omega^{-1} = \emptyset$, contrary to hypothesis.

Let ν be the projection of $X = Y \times \mathcal{R}$ onto $\mathcal{R}$. Since Y is compact, f is bounded on Y and there exists $a \in \mathcal{R}^+$ such that $x \in X$ implies $| x\nu - x\varphi\nu | < a$. Let $b = \sup [r \mid r \in I_u]$ and let $c \in R$ with $c > b + Ma$. Then $x \in X$ with $x\nu > c$ implies $x\varphi^i\nu > b$, $i = 1, 2, \cdots, M$.

There exists $L \in \mathcal{I}^+$ such that $n \in \mathcal{I}^+$ implies that at least one of the points $x_-\varphi^{n+1}, \cdots, x_-\varphi^{n+L}$ satisfies the condition $x\nu > c$. For otherwise there exist sequences $n_1, n_2, \cdots$ and $L_1, L_2, \cdots$ of positive integers with $\lim_{i \to +\infty} L_i = +\infty$ and such that $i \in \mathcal{I}^+$ implies that $x_-\varphi^j\nu \leqq c$, $j = n_i + 1, \cdots, n_i + L_i$. Since $x_- \in \Omega^-$ there exists $d \in \mathcal{R}$ such that $n \in \mathcal{I}^+$ implies $x_-\varphi^n\nu > d$. The sequence $(x_-\varphi^{n_i+1} \mid i \in \mathcal{I}^+)$ contains a subsequence converging to a point x_0 of X. The positive semiorbit of x_0 is bounded and it would follow from 14.11 that (3) is not valid, contrary to hypothesis.

Let $e = \inf [r \mid r \in I_u]$. Since $x_+ \in \Omega^{-\prime}$ there exists $t \in \mathcal{I}^+$ such that $x_+\varphi^{t+i}\nu < e$, $i = 1, 2, \cdots, L + M$. Of the points $x_-\varphi^{t+1}, x_-\varphi^{t+2}, \cdots, x_-\varphi^{t+1}$, at least one, $x_-\varphi^{t+j}$, is such that $x_-\varphi^{t+j}\nu > c$. Thus $x_-\varphi^{t+k}\nu > b$ and $x_+\varphi^{t+k}\nu < e$, $k = j + 1, \cdots, j + M$. There exists an integer p with $t + j + 1 \leqq p \leqq t + j + M$ such that $Y_v\theta^p \subset Y_u$. Consider the connected set $C\varphi^p \subset Y_u \times \mathcal{R}$. It contains $x_-\varphi^p$ and $x_+\varphi^p$. Since $x_-\varphi^p\nu > b$ and $x_+\varphi^p\nu < e$, it follows that $C\varphi^p$ must meet $Y_u \times I_u$ and thus U. We infer that $U \cap \bigcup_{n \in \mathcal{I}} V\varphi^n \neq \emptyset$, which implies (1).

The proof that (4) implies (1) is similar and will be omitted.

Assume (5). We prove (3). Let $x \in A^{-\prime} \cap \Omega^{-\prime}$. By 14.08, $A^+ \neq \emptyset \neq \Omega^-$. Let $x^* \in \Omega^-$. Then $x^* \in \Omega^{+\prime}$. For otherwise it follows from 14.11 that $A^{-1} = \emptyset$, contrary to hypothesis. This proves (3).

Similarly, (6) implies (4).

Assume (6). We prove (7). It is known (cf. Oxtoby [3]) that

$$\lim_{n \to +\infty} \frac{1}{n} \sum_{p=0}^{n-1} f(y\theta^p) = \int_Y f(y)\, d\mu(y) = \lim_{n \to +\infty} \frac{1}{n} \sum_{p=1}^{n} f(y\theta^{-p}), \qquad y \in Y.$$

Thus, if $\int_Y f(y)\, d\mu(y) > 0$, then $\omega^+ = X = \alpha^-$ and if $\int_Y f(y)\, d\mu(y) < 0$, then $\omega^- = X = \alpha^+$. In either case (6) is not valid and we infer that $\int_Y f(y)\, d\mu(y) = 0$. Since $x = (y, r) \in X$ and $n \in \mathscr{I}^+$ implies $x\varphi^n = (y\theta^n, \sum_{p=0}^{n-1} f(y\theta^p))$, it follows from (6) that $\sum_{p=0}^{n-1} (y\theta^p)$ is not bounded on $Y \times \mathscr{I}^+$. The proof that (6) implies (7) is completed.

Assume (7). We prove (5). Since $\int_Y f(y)\, d\mu(y) = 0$, it follows that

$$\int_Y \left(\sum_{p=0}^{n-1} f(y\theta^p) \right) d\mu(y) = 0, \qquad n \in \mathscr{I}^+,$$

and thus, corresponding to $n \in \mathscr{I}^+$ there exists $y_n \in Y$ such that $\sum_{p=0}^{n-1} f(y_n\theta^p) = 0$. Let $x_n = (y_n, 0)$. Then $x_n\nu = 0$ and $x_n\varphi^n\nu = 0$. Corresponding to $m \in \mathscr{I}^+$ there exists $n(m) \in \mathscr{I}^+$ such that not all the points $x_{n(m)}\varphi^j$, $j = 0, 1, \cdots, n(m)$, satisfy the condition $| x\nu | \leqq m$. For otherwise there would exist a bounded orbit under φ, and from 14.11 it would follow that $\sum_{p=0}^{n-1} f(y\theta^p)$ is bounded on $Y \times \mathscr{I}^+$, contrary to hypothesis. By 14.09, either $B^+ \neq \emptyset$ or $B^- \neq \emptyset$. If $B^+ \neq \emptyset$ and $x \in B^+$, it follows from 14.11 that $x \in A^{-1} \cap \Omega^{-\prime}$ and thus (5) is valid. If $B^- \neq \emptyset$ and $x \in B^-$, it follows from 14.11 that $x \in A^{+1} \cap \Omega^{+\prime}$ and thus (6) is valid. But it has been shown that (6) implies (5), and thus, in either case, (5) is valid.

The proof of the theorem is completed.

14.14. Examples of cylinder flows. It follows from 14.11 that it is easy to construct nontrivial examples of cylinder homeomorphisms which are pointwise almost periodic. Using the notation of 14.11, we choose $g \in F(Y)$ and define $f \in F(Y)$ by $f(y) = g(y\theta) - g(y)$, $y \in Y$. Then $\varphi = \varphi(Y, f, \theta)$ is pointwise almost periodic.

It is more difficult to construct examples of transitive cylinder flows. The following method yields such examples.

Let Y be a compact connected separable abelian (additive) topological group. Then Y is monothetic (cf. Halmos and Samelson [1]). Let y^* be a generator of Y; that is, $Y = [ny^* \mid n \in \mathscr{I}]^-$. Let $\theta : Y \to Y$ be defined by $y\theta = y + y^*$, $y \in Y$. Then Y is an almost periodic minimal orbit-closure under θ.

Let C be the unit circle $z\bar{z} = 1$ of the complex plane Z. Let $\chi : Y \to C$ be a character of Y such that $\chi(y^*) = e^{i\beta} \neq 1$. Since Y is connected, β/π is irrational.

Let $0 < n_1 < n_2 < \cdots$ be a sequence of integers such that $\sum_{k=1}^{\infty} | \chi(n_k y^*) - 1 | = \sum_{k=1}^{\infty} | e^{in_k\beta} - 1 |$ is convergent. Since β/π is irrational, such a sequence exists. Let $(a_n \mid n \in \mathscr{I})$ be defined by:

$$\begin{cases} a_n = 0 \text{ unless } n \in [n_k \mid k \in \mathrm{I}^+] \text{ or } n \in [-n_k \mid k \in \mathscr{I}^+], \\ a_n = a_{-n_k} = | e^{in_k\beta} - 1 |, \qquad k \in \mathscr{I}^+. \end{cases}$$

The series $\sum_{-\infty}^{+\infty} a_n\, \chi(ny)$ is absolutely and uniformly convergent on Y. Let $f : Y \to Z$ be defined by $f(y) = \sum_{-\infty}^{+\infty} a_n x(ny)$, $y \in Y$. Then f is continuous on Y and since $\chi(-y) = \bar{\chi}(y)$, $y \in Y$, it follows that $f(y) \in \mathcal{R}$, $y \in Y$. By the orthogonality property of characters, $n \in \mathcal{I}$, $n \neq 0$, implies $\int_Y \chi(ny)\, d\mu(y) = 0$, where μ is the normalized Haar measure on Y, and thus $\int_Y f(y)\, d\mu(y) = 0$.

Now suppose that there exists $g \in F(Y)$ such that

$$\text{(A)} \qquad f(y) = g(y + y^*) - g(y), \qquad y \in Y.$$

Let $b_n = \int_Y g(y)\bar{\chi}(ny)\, d\mu(y)$, $n \in \mathcal{I}$. Then $\sum_{-\infty}^{+\infty} b_k \bar{b}_k < \infty$. But a simple computation shows that $a_n = b_n(e^{in\beta} - 1)$, $n \in \mathcal{I}$. Thus

$$| b_{n_k} | = | a_{n_k} | \, | e^{in_k\beta} - 1 |^{-1} = 1, \qquad k \in \mathcal{I}^+,$$

and hence $\sum_{-\infty}^{+\infty} b_k \bar{b}_k = \infty$. We infer that there cannot exist $g \in F(Y)$ such that (A) is valid.

Let Y be also locally connected and let $\varphi = \varphi(Y, f, \theta)$. It follows from 14.11 that $\sum_{p=0}^{n-1} f(y\theta^p)$ is not bounded on $Y \times \mathcal{I}^+$ and thus, by 14.13, the discrete flow generated by φ is transitive.

14.15. Remark. The *dyadic tree* is a dendrite whose endpoints form a Cantor discontinuum and whose branch points are all of order three. There exists a homeomorphism φ of the dyadic tree X onto X such that:

(1) φ is regularly almost periodic on X.
(2) φ is periodic at every cut point of X.
(3) The set of all endpoints of X is a minimal orbit-closure under φ.

14.16. Remark. The remainder of this section is devoted to the construction of a compact, connected plane set which is minimal under a homeomorphism and which is locally connected at some points and not locally connected at other points.

14.17. Remark. Let f be a continuous real-valued function on a dense subset of a real interval such that the closure of (the graph of) f in the plane is compact, connected and locally connected. Then f is uniformly continuous. (Since $\bar{f}$ is compact, it is enough to show that the relation $\bar{f}$ is single-valued. This may be done by use of the arcwise connectedness theorem.)

14.18. Definition. Let $A_0 = [2k\pi \mid k \in \mathcal{I}]$, let $X_0 = \mathcal{R} - A_0$ and let $f_0 : X_0 \to \mathcal{R}$ be defined by:

$$\begin{cases} f_0(x) = \sin \pi^2 \, | x |^{-1}, & -\pi \leqq x \leqq \pi, \quad x \neq 0, \\ f_0(x + 2\pi) = f_0(x), & x \in X_0 . \end{cases}$$

For $n \in \mathcal{I}$, $n \neq 0$, let $A_n = [2k\pi + n \mid k \in \mathcal{I}]$, let $X_n = \mathcal{R} - A_n$ and let $f_n : X_n \to \mathcal{R}$ be defined by $f_n(x) = f_0(x - n)$, $x \in X_n$.

Let $X = \mathcal{R} - \bigcup_{n \in \mathcal{I}} A_n = \bigcap_{n \in \mathcal{I}} X_n$ and let $f : X \to \mathcal{R}$ be defined by:

$$f(x) = 10 + \sum_{-\infty}^{+\infty} 2^{-|n|} f_n(x), \qquad x \in X.$$

14.19. REMARK. We adopt the notation of 14.18. The following statements are valid:

(1) f is continuous on X.

(2) If I is any open interval of $\mathcal{R}$, then f is not uniformly continuous on $X \cap I$.

(3) f is δ-chained for every $\delta \in \mathcal{R}^+$.

14.20. LEMMA. *We adopt the notation of* 14.18. *Let* $N = [(x, f(x)) \mid x \in X]$. *We consider* N *as a subspace of the product space* $\mathcal{R} \times \mathcal{R}$. *Let* $\psi : N \to N$ *be defined by* $(x, f(x))\psi = (x + 1, f(x + 1))$, $x \in X$. *Then* ψ *and* ψ^{-1} *are uniformly continuous homeomorphisms of* N *onto* N.

PROOF. It is clear that ψ is a one-to-one transformation of N onto N. That ψ and ψ^{-1} are continuous on N follows from the continuity of f on X. Thus ψ is a homeomorphism of N onto N. We prove that ψ is uniformly continuous on N; the proof that ψ^{-1} is uniformly continuous is similar.

Suppose that ψ is not uniformly continuous on N. Then there exists $\epsilon \in \mathcal{R}^+$ such that corresponding to $\delta \in \mathcal{R}^+$ there exist $x, x' \in X$ such that $|x - x'| < \delta$, $|f(x) - f(x')| < \delta$ and $|f(x + 1) - f(x' + 1)| > \epsilon$. Let $(\delta_n \mid n \in \mathcal{I}^+)$ be a sequence of positive real numbers such that $\delta_1 > \delta_2 > \cdots$ and $\lim_{n\to+\infty} \delta_n = 0$. Then there exists a sequence of pairs $((x_n, x_n') \mid n \in \mathcal{I}^+)$ such that $x_n, x_n' \in X$, $|x_n - x_n'| < \delta_n$, $|f(x_n) - f(x_n')| < \delta_n$, $|f(x_n + 1) - f(x_n' + 1)| > \epsilon$, for all $n \in \mathcal{I}^+$.

We can assume that $\lim_{n\to+\infty} x_n = \bar{x} = \lim_{n\to+\infty} x_n'$, where $\bar{x} \in \mathcal{R}$. Since $\liminf_{n\to+\infty} |f(x_n + 1) - f(x_n' + 1)| \geqq \epsilon$ it follows that $\bar{x} + 1 \notin X$ and there exist $k, m \in \mathcal{I}$ such that $\bar{x} + 1 = 2k\pi + m$.

If $n \in \mathcal{I}^+$, then

$$\epsilon < |f(x_n + 1) - f(x_n' + 1)| = \left| \sum_{-\infty}^{+\infty} 2^{-|p|}[f_p(x_n + 1) - f_p(x_n' + 1)] \right|.$$

There exists $M \in \mathcal{I}^+$ such that $M > m$ and

$$\left| \sum_{-\infty}^{-M-1} 2^{-|p|}[f_p(x_n + 1) - f_p(x_n' + 1)] \right| < \epsilon/3$$

and

$$\left| \sum_{M+1}^{+\infty} 2^{-|p|}[f_p(x_n + 1) - f_p(x_n' + 1)] \right| < \epsilon/3, \qquad n \in \mathcal{I}^+.$$

Thus

$$\left| \sum_{-M}^{M} 2^{-|p|}[f_p(x_n + 1) - f_p(x_n' + 1)] \right| > \epsilon/3, \qquad n \in \mathcal{I}^+.$$

Since $p \in \mathcal{I}$, $p \neq m$, implies

$$\lim_{n\to+\infty} |f_p(x_n + 1) - f_p(x_n' + 1)| = 0,$$

there exists $P \in \mathcal{I}^+$ such that $n \in \mathcal{I}^+$, $n > P$, implies $| 2^{-|m|}[f_m(x_n + 1) - f_m(x'_n + 1)] | > \epsilon/4$. Since $f_m(x_n + 1) = f_{m-1}(x_n)$, $n \in \mathcal{I}^+$, we have

$$| 2^{-|m|}[f_{m-1}(x_n) - f_{m-1}(x'_n)] | > \epsilon/4$$

provided $n > P$, $n \in \mathcal{I}^+$.

Since $n \in \mathcal{I}^+$ implies

$$| f(x_n) - f(x'_n) | = \left| \sum_{-\infty}^{+\infty} 2^{-|p|}[f_p(x_n) - f_p(x'_n)] \right| < \delta_n ,$$

there exists $Q \in \mathcal{I}^+$, $Q > m$, such that

$$\left| \sum_{-Q}^{Q} 2^{-|p|}[f_p(x_n) - f_p(x'_n)] \right| < \delta_n + \epsilon/10 \qquad (n \in \mathcal{I}^+).$$

Since $p \in \mathcal{I}$, $p \neq m - 1$, implies

$$\lim_{n \to +\infty} | f_p(x_n) - f_p(x'_n) | = 0,$$

there exists $S \in \mathcal{I}^+$ such that $n > S$ implies

$$| 2^{-|m-1|}[f_{m-1}(x_n) - f_{m-1}(x'_n)] | < \delta_n + \epsilon/9.$$

Let $t \in \mathcal{I}^+$ with $t > P$ such that

$$| 2^{-|m-1|}[f_{m-1}(x_t) - f_{m-1}(x'_t)] | < \epsilon/8,$$

whence

$$\frac{\epsilon}{8} > 2^{-|m-1|} \, | f_{m-1}(x_t) - f_{m-1}(x'_t) | > \frac{2^{-|m-1|}}{2^{-|m|}} \frac{\epsilon}{4} \geqq \frac{\epsilon}{8}.$$

From this contradiction we infer the validity of the lemma.

14.21. Definition. We adopt the notation of 14.18. Let M be the graph of $r = f(\theta)$, $\theta \in X$, in polar coordinates. Let $\varphi : M \to M$ be defined by $(\theta, f(\theta))\varphi = (\theta + 1, f(\theta + 1))$, $\theta \in X$.

14.22. Theorem. *The transformation $\varphi : M \to M$ is a uniformly continuous homeomorphism of M onto M, φ^{-1} is uniformly continuous on M, φ and φ^{-1} are pointwise almost periodic, and M is a minimal orbit-closure under φ.*

Proof. It is obvious that φ is a homeomorphism of M onto M, that φ and φ^{-1} are pointwise almost periodic and that M is a minimal orbit-closure under φ. The uniform continuity of φ and φ^{-1} follows from 14.20.

14.23. Remark. Let X be a compact metric space, let Y be a nonvacuous subset of X and let φ be a pointwise almost periodic homeomorphism of Y onto Y such that φ and φ^{-1} are uniformly continuous on Y and such that $\overline{Y}$ is a minimal orbit-closure under φ. Then there exists a homeomorphism $\hat{\varphi}$ of $\overline{Y}$ onto $\overline{Y}$ such that $\hat{\varphi} \mid Y = \varphi$ and $\overline{Y}$ is a minimal orbit-closure under $\hat{\varphi}$.

14.24. Theorem. *Let M be the plane set defined in* 14.21. *Then $\overline{M}$ is compact, connected, and locally connected at some points but not locally connected at other points. There exists a homeomorphism η of $\overline{M}$ onto $\overline{M}$ such that $\overline{M}$ is a minimal orbit-closure under η.*

Proof. Clearly M is locally connected at each of the points $(\theta, f(\theta))$, $\theta \in X$. To complete the proof, use 14.17, 14.19(3), 14.20, 14.22, 14.23.

14.25. Notes and references.

(14.01) Cylinder homeomorphisms with Y a circle were considered by A. S. Besicovitch [1, 2], who constructed transitive models of cylinder flows.

(14.15) Cf. Zippin [1], pp. 196–197 and Gottschalk [4].

(14.16–14.24) This example of a minimal set is due to F. B. Jones (Personal communication).

APPENDIX[2]

MINIMAL SETS: AN INTRODUCTION TO TOPOLOGICAL DYNAMICS[1]

W. H. GOTTSCHALK

The notion of minimal set is centrally located in topological dynamics. Topological dynamics may be defined as the study of transformation groups with respect to those properties, wholly or largely topological in nature, whose prototype occurred in classical dynamics.

Henri Poincaré was the first to introduce topological notions and methods in dynamics, that is, the study of ordinary differential equations. G. D. Birkhoff was the first to undertake the systematic development of topological dynamics, indicating its essentially abstract character and making fundamental contributions. Birkhoff's first paper on the subject appeared in 1912 [**6**, pp. 654–672]; Birkhoff's paper contains the first definition of minimal sets, some theorems about them, and some examples of them. Most of Birkhoff's work in topological dynamics from the point of view of general theory is to be found in Chapter 7 of his Colloquium volume *Dynamical systems* published in 1927 [**5**]. The Colloquium volume *Analytical topology* by G. T. Whyburn [**27**], published in 1942, contains related developments in its Chapter 12. The Russian book, *Qualitative theory of differential equations* by Nemyckiĭ and Stepanov [**22**], first edition in 1947 and second edition in 1949, contains a chapter devoted to topological dynamics; an English translation of this book has recently been announced by Princeton Press. The treatments of topological dynamics in the above books are all from the points of view of a single transformation or a one-parameter group of transformations. The Colloquium volume *Topological dynamics* by Hedlund and myself [**14**], published in 1955, is concerned mainly with general transformation groups.

A *topological transformation group*, or *transformation group* for short, is defined to be an ordered triple (X, T, π) such that the following axioms are satisfied:

(A0) (Stipulative axiom). X is a topological space, called *the phase space*; T is a topological group, called the *phase group*; and π is a map of the cartesian product $X \times T$ into X, called the *phase map*.

An address delivered before the Washington meeting of the Society on October 26, 1957, by invitation of the Committee to Select Hour Speakers for Eastern Sectional Meetings; received by the editors April 10, 1958.

[1] This address was prepared while the author was under contract No. AF 18(600)-1116 of the Air Force Office of Scientific Research. Reproduction in whole or in part is permitted for any purpose of the United States Government.

[2] **Reprinted from the Bulletin of the American Mathematical Society, Vol. 64, No. 6, November, 1958, pp. 336-351.**

[Let us for the moment use the multiplicative notation for the phase group T, let e denote the identity element of T, and let the value of π at the point (x, t) of $X \times T$ be denoted by xt.]

(A1) (IDENTITY AXIOM). $xe = x$ for all $x \in X$.

(A2) (HOMOMORPHISM AXIOM). $(xt)s = x(ts)$ for all $x \in X$ and all $t, s \in T$.

(A3) (CONTINUITY AXIOM). π is continuous.

Consider a given transformation group (X, T, π). The phase map $(x, t) \rightarrow xt$ determines two kinds of maps when one of the variables x, t is replaced by a constant. Thus, for fixed $t \in T$, the map $x \rightarrow xt$ is a homeomorphism π^t of the phase space X onto itself which is called a *transition*. Again, for fixed $x \in X$, the map $t \rightarrow xt$ is a continuous map π_x of the phase group T into the phase space X which is called a *motion*. The set of all transitions is a group of homeomorphisms of the phase space X onto itself such that the transition π^e induced by the identity element e of T is the identity homeomorphism of the phase space X and such that the map $t \rightarrow \pi^t$ is a group homomorphism. Conversely, a group of homeomorphisms of X onto itself, suitably topologized, gives rise to a transformation group.

An *intrinsic* property of the transformation group (X, T, π) is defined to be a property which is describable in terms of the topology of the phase space X, the topology and group structure of the phase group T, and the phase map π. Topological dynamics is concerned with intrinsic properties of transformation groups with particular reference to those properties which first arose in classical dynamics.

Among the examples which have been studied in the past, the most frequently occurring phase groups are the additive group $\mathcal{I}$ of integers with the discrete topology and the additive group $\mathcal{R}$ of real numbers with its usual topology. In either case, the transformation group is called a *flow*. If the phase group T is $\mathcal{I}$, then the transformation group is called a *discrete* flow. If the phase group T is $\mathcal{R}$, then the transformation group is called a *continuous* flow. Here the word "continuous" apparently refers to the "real number continuum".

To see how transformation groups in the role of continuous flows appeared in classical dynamics, consider a system of n first order ordinary differential equations of the form

$$\frac{dx_i}{dt} = f_i(x_1, \cdots, x_n) \qquad (i = 1, \cdots, n)$$

where the functions f_i may be defined on a region X of euclidean n-

space or they may be defined locally throughout an n-dimensional differentiable manifold X. Impose enough conditions to insure the existence, uniqueness, and continuity of solutions for all real values of the "time" t and through all points of X. Then there exists exactly one continuous flow in X whose motions are the solutions of the system of differential equations. To illustrate with a very simple example, the continuous flow in the plane determined by the system

$$\frac{dx_1}{dt} = x_1, \qquad \frac{dx_2}{dt} = -x_2$$

has as phase map $\pi((x_1, x_2), t) = (x_1 e^t, x_2 e^{-t})$.

A *dynamical system* may be defined as a continuous flow in a differentiable manifold induced by an autonomous system of first order ordinary differential equations, as above. Of course, every dynamical system is a continuous flow. To what extent and under what conditions a continuous flow is a dynamical system appears to be a largely unsolved problem. Let me say that this usage of the term "dynamical system" is not at all universal. More likely than not it is often used as synonymous with "continuous flow" or even "discrete flow."

The two kinds of flows, discrete and continuous, are closely related. For example, a continuous flow determines many discrete flows by taking cyclic subgroups of $\mathcal{R}$. Conversely, a discrete flow determines a continuous flow when the phase space X is extended to the cartesian product of X and the closed unit interval, and the bases of the cylinder thus obtained are identified according to the transition induced by the integer 1, that is, $(x, 1)$ and $(x1, 0)$ are identified. The problem of imbedding a discrete flow in a continuous flow without alteration of the phase space appears to be largely unsolved.

A discrete flow determines and is determined by the transition induced by the integer 1. Thus, we may redefine a discrete flow more economically as an ordered pair (X, ϕ) where X is a topological space and ϕ is a homeomorphism of X onto X. For the sake of simplicity, I wish to confine most of my remaining remarks to flows, and in particular to discrete flows. Moreover, the phase space will be assumed to be compact metric for the same reason. It is to be pointed out that the present theory of topological dynamics contains results on transformation groups acting upon topological spaces and uniform spaces which specialize to give most of the known facts about flows on metric spaces.

Let X be a nonvacuous compact metric space with metric ρ and let ϕ be a homeomorphism of X onto X.

The phase space X is nonvacuous closed and invariant, that is, $X\phi = X$. In case X has these properties minimally it is said that X is a *minimal set*. The definition may be phrased for a subset of X. A subset M of X is said to be *minimal under* ϕ provided that M is nonvacuous closed and invariant, that is $M\phi = M$, and no proper subset of M has all these properties. Minimal sets may be characterized in various ways. If $x \in X$, then the *orbit of* x, denoted $O(x)$, is defined to be $\{x\phi^n \mid n \in \mathcal{I}\}$ where $\mathcal{I}$ is the set of all integers; and the *orbit-closure of* x, denoted $\overline{O}(x)$, is defined to be the closure of the orbit $O(x)$ of x. A subset M of X is minimal if and only if M is nonvacuous and M is the orbit-closure of each of its points.

By the axiom of choice, there always exists at least one minimal subset of X [**14**, p. 15].

Now recursive properties enter into consideration. A point x of the phase space is said to be *almost periodic under* ϕ and ϕ is said to be *almost periodic at* x provided that if U is a neighborhood of x, then there exists a relatively dense subset A of $\mathcal{I}$ such that $x\phi^n \in U$ for all $n \in A$. A subset A of $\mathcal{I}$ is called *relatively dense* in case the gaps of A are bounded. Another way of saying the same thing is that $\mathcal{I} = A + K = \{a + k \mid a \in A \& k \in K\}$ for some finite subset K of $\mathcal{I}$. We may call the set K a *bond* of the set A. A periodic point is almost periodic, but not conversely. It has been remarked that a periodic point returns to itself every hour on the hour; but an almost periodic point returns to a neighborhood every hour within the hour.

The basic connection between minimal sets and almost periodic points is this: The orbit-closure of a point x is minimal if and only if the point x is almost periodic [**14**, p. 31]. Consequently, every point of a minimal set is almost periodic, and there always exists at least one almost periodic point. This, of course, is in strong contrast to the situation with respect to periodic points.

Since different minimal subsets of X are necessarily disjoint, we may conclude that the class of all orbit-closures is a partition of the phase space if and only if ϕ is pointwise almost periodic, that is to say, ϕ is almost periodic at each point of X. The partition of orbit-closures, when it exists, is necessarily star-open in the sense that the saturation or star of every open set is again open. In a compact metric space, which we are here considering, this is equivalent to lower semi-continuity in another terminology. It is however not always the case that the partition of orbit-closures is star-closed in the sense that the star of every closed set is itself closed, or equivalently, upper semi-continuous. This condition may be characterized by the following recursive property. The map ϕ is said to be *weakly almost periodic on*

X and the phase space X is said to be be *weakly almost periodic under* ϕ provided that if ϵ is a positive real number, then there exists a finite subset K_ϵ of $\mathcal{I}$ such that $x \in X$ implies the existence of a (necessarily relatively dense) subset A_x of $\mathcal{I}$ such that $\mathcal{I} = A_x + K_\epsilon$ and $x\phi^n \in N_\epsilon(x)$ for all $n \in A_x$. Here the bond K_ϵ depends only upon ϵ, but the individual sets of return A_x depend also upon x. To summarize, the class of all orbit-closures is a star-closed partition of the phase space if and only if the map ϕ is weakly almost periodic [**14**, p. 34]. In particular, ϕ is weakly almost periodic on each minimal set.

We are now in the middle of the spectrum of recursive properties. A weaker recursive property than almost periodic may be mentioned first of all. A point x of the phase space is said to be *recurrent* provided that if U is a neighborhood of x, then there exists an extensive subset A of $\mathcal{I}$ such that $x\phi^n \in U$ for all $n \in A$. Here "extensive" means containing a sequence diverging to $-\infty$ and a sequence diverging to $+\infty$. In other words a point is recurrent provided that it returns (or recurs) to a neighborhood infinitely often in the past and infinitely often in the future. The term recurrent as it is here used coincides with its meaning in the Poincaré Recurrence Theorem (1899), which seems to be one of the earliest appearances in the literature of a theorem on a recursive property different from periodic. In general, pointwise recurrent does not imply pointwise almost periodic [**12**]. However, if the phase space is also zero-dimensional, then pointwise recurrent implies weakly almost periodic [**14**, p. 65].

Corresponding to any recursive property there is also the property of regionally recursive. For example, ϕ is *regionally recurrent* provided that if $x \in X$ and if U is a neighborhood of x, then there exists an extensive subset A of $\mathcal{I}$ such that $U \cap U\phi^n \neq \emptyset$ for every $n \in A$. The *center* of a discrete flow (X, ϕ) is defined to the greatest invariant subset on which ϕ is regionally recurrent. The center is also characterized as the closure of the set of recurrent points. Since every almost periodic point is also recurrent, the center is necessarily nonvacuous. The center has the property that the relative sojourn of every point of the phase space in a neighborhood of the center is equal to unity. If the flow has an invariant measure which is positive and finite for nonvacuous open sets, then the flow is regionally recurrent and the center is the entire phase space. This is the case for conservative dynamical systems [**14**; **22**].

Let us examine a few stronger recursive properties. The map ϕ is said to be *almost periodic on* X and the phase space X is said to be *almost periodic under* ϕ provided that if $\epsilon > 0$, then there exists a relatively dense subset A of $\mathcal{I}$ such that $x\phi^n \in N_\epsilon(x)$ for all $x \in X$ and all $n \in A$. Almost periodic implies weakly almost periodic which in turn implies pointwise almost periodic; no converse holds. The almost periodicity of the map ϕ may be characterized as follows. These four statements are pairwise equivalent: (1) ϕ is almost periodic; (2) the set of powers of ϕ, namely $\{\phi^n | n \in \mathcal{I}\}$, is equicontinuous; (3) the set of powers of ϕ has compact closure where the ambient space is the group of all homeomorphisms of X onto X with the usual topology; (4) there exists a compatible metric of X which makes ϕ an isometry [**14**, p. 37].

A still stronger recursive property is regularly almost periodic (G. T. Whyburn 1942; P. A. Smith 1941 related) [**27**; **24**]. This is like almost periodic except that the set of return A is taken to be a nontrivial subgroup of $\mathcal{I}$. We have then the properties: regularly almost periodic point, pointwise regularly almost periodic map, and regularly almost periodic map. We mention a few results in this context. The map ϕ is regularly almost periodic if and only if ϕ is both pointwise regularly almost periodic and weakly almost periodic. If ϕ is pointwise regularly almost periodic, then every orbit-closure is a zero-dimensional regularly almost periodic minimal set [**14**, p. 49f.]. That a regularly almost periodic homeomorphism on a manifold is necessarily periodic has been conjectured by P. A. Smith (1941) [**24**], but so far has been proved only for the two-dimensional case [for example, **14**, p. 56].

By definition, an *inheritance* theorem is a theorem of the form "ϕ has property P if and only if ϕ^n has the property P" where n is a preassigned nonzero integer. The inheritance theorem holds for recurrent and almost periodic points as well as for certain other recursive properties of a point. Actually, a very general inheritance theorem can be proved which yields most known results for recursive points [**14**, p. 26]. No general inheritance theorem is known for such properties as regionally recurrent and weakly almost periodic although they too inherit, at least under certain conditions [**14**, pp. 67 and 35].

In this connection we may mention the fact that a connected minimal set under ϕ is also minimal under ϕ^n $(n \neq 0)$ [**14**, p. 16].

Two discrete flows (X, ϕ) and (Y, ψ) are said to be *isomorphic* provided there exists a homeomorphism h of X onto Y such that $\phi h = h\psi$.

Let us consider now the problems: (1) (Construction problem.) To construct all minimal sets systematically; (2) (Classification problem.) To classify all minimal sets according to isomorphism type. At the present, only partial answers can be given to these questions, even for discrete flows on compact metric spaces.

In the spectrum of isomorphism types of minimal sets, those which are almost periodic seem to constitute an extreme case, for the following reason. A *monothetic group* is by definition a compact group which contains a dense cyclic subgroup, a generator of a dense cyclic subgroup being called a generator of the monothetic group. Now the almost periodic minimal sets under discrete flows are coextensive with the monothetic groups in the following sense. If X is a monothetic group with generator a, then X is a minimal set under the homeomorphism ϕ of X onto X defined by $x \to xa$. Conversely, if (X, ϕ) is an almost periodic minimal set, then there exists a group structure in X which makes X a monothetic group and such that the map ϕ is translation by a generator a, that is $x\phi = xa$ for all $x \in X$ [**14**, p. 39]. Since the monothetic groups can be constructed in their entirety as the character groups of the subgroups of the discrete circle group (Anzai and Kakutani 1943) [**2**], the construction problem for almost periodic minimal sets under discrete flows has a reasonably definite answer. It is not clear to me whether the corresponding classification problem can also be answered by the present theory of topological groups. It is known (Halmos and Samelson 1942; Anzai and Kakutani 1943) [**15**; **2**] that every compact connected separable abelian group is monothetic. For example, the n-toral groups K^n where n is a positive integer and the infinite toral group $K^{\mathfrak{g}}$ are monothetic.

The n-adic groups are monothetic and are topologically the Cantor discontinuum. Let us describe geometrically a minimal set given by the dyadic group and a generator. A similar construction is available for the general case.

If E is the disjoint union of segments (meaning closed line segments) $E_1, \cdots, E_r$, let E^* denote the union of segments obtained by deleting the open middle third of each of the segments $E_1, \cdots, E_r$. Let S be a segment of length 1. Define $X_0, X_1, \cdots$ inductively as follows: $X_0 = S$, $X_{n+1} = X_n^*$ $(n = 0, 1, 2, \cdots)$. Define $X = \bigcap_{n=0}^{+\infty} X_n$. The space X is the Cantor discontinuum. Denote the 2^n disjoint segments which make up X_n $(n = 0, 1, 2, \cdots)$ by $S(n, m)$ $(0 \leqq m \leqq 2^n - 1)$ where $S(0, 0) = S$, $S(n+1, m)$ is the initial third of $S(n, m)$, and $S(n+1, m+2^{n-1})$ is the terminal third of $S(n, m)$. For each nonnegative integer n permute the segments $S(n, m)$ $(0 \leqq m \leqq 2^{n-1})$ cyclicly according to the second coordinate m. These permutations induce a homeomorphism ϕ of X onto itself. More exactly, if $x \in X$, then for each nonnegative integer n we have that $x \in S(n, m_n)$ for exactly one integer m_n with $0 \leqq m_n \leqq 2^n - 1$ and $x\phi = \bigcap_{n=0}^{+\infty} S(n, m_n + 1 \pmod{2^n}))$. It follows that (X, ϕ) is a regularly almost periodic minimal set which we may call the dyadic minimal set. To see the group structure of X, draw the horizontal segments $S(1, 0)$ and $S(1, 1)$ under the horizontal segment $S(0, 0)$, draw downward arrows to $S(1, 0)$ and $S(1, 1)$ from $S(0, 0)$, and label the arrows 0 and 1 from left to right. Continue this process. A point of X is uniquely represented by a downward chain of arrows and therefore by a sequence of 0's and 1's. The sequences are added coordinatewise with (possibly infinite) carry-over. The generator of the dyadic group X which corresponds to ϕ is given by the sequence $(1, 0, 0, 0, \cdots)$. The n-adic minimal sets are all regularly almost periodic.

In the case of the circle (1-sphere) as a minimal set under a discrete flow, the isomorphism type is characterized by an irrational number r between 0 and 1, the Poincaré rotation number, and the minimal set is isomorphic to the spin of a circle through an angle which is $r\pi$ [for example, **17**]. Actually, the discrete flows on a circle appear to be presently unclassified. Let us look at one. Let $\cdots, E_{-1}, E_0, E_1, \cdots$ be a disjoint bisequence of closed arcs on a circle K such that $\bigcup_{n\in\mathcal{g}} E_n$ is dense in K and such that the cyclic order of $\cdots, E_{-1}, E_0, E_1, \cdots$ on K agrees with the cyclic order of $\cdots, xr^{-1}, x, xr, \cdots$ on a circle when the circle is rotated through one radian, x being a point of the circle and r denoting the rotation. Map E_n $(n\in\mathcal{g})$ homeomorphically onto E_{n+1} in the evident manner. This defines a homeomorphism of $\bigcup_{n\in\mathcal{g}} E_n$ onto itself. Since both this homomorphism and its inverse are uniformly continuous, it can be extended uniquely to a homeomorphism ϕ of K onto K. The complement of $\bigcup_{n\in\mathcal{g}} \operatorname{int} E_n$ is a minimal set under ϕ which is the Cantor discontinuum and which is not almost periodic. The endpoints of the E_n $(n\in\mathcal{g})$ constitute a pair of doubly asymptotic orbits, that is $\rho(x\phi^n, y\phi^n)$ goes to zero with $1/n$ where x and y are the endpoints of the arc E_0.

The almost periodic minimal sets are homogeneous since their phase spaces are groups. Is every minimal set homogeneous? The answer is no (Floyd 1949) [**11**]. Floyd's example may be described in geometric language (University of Pennsylvania dissertation of Joseph Auslander 1957) [**3**].

If E is a box $A\times B$ where A and B are segments, define E^* to be $(A_1\times B_1)\cup(A_3\times B)\cup(A_5\times B_2)$ where $A_1, \cdots, A_5$ are the consecutive equal fifths of A and B_1, B_2 are the consecutive equal halves of B. If D is a disjoint union of boxes, define D^* to be the union of the E^* where E ranges over the boxes in D. Start with the unit box X_0 and define

$$X_{n+1} = X_n^* \ (n = 0, 1, 2, \cdots), \qquad X = \bigcap_{n=0}^{+\infty} X_n.$$

The boxes in X_n ($n=0, 1, 2, \cdots$) are permuted just like the segments are permuted in the construction of the triadic minimal set. For example, the first three stages have boxes numbered as follows from left to right: 0; 0, 1, 2: 0, 3, 6, 1, 4, 7, 3, 5, 8. The process gives rise to a homeomorphism ϕ of X onto X. It is seen that X consists of point-components and segments, X is zero-dimensional at point-components, X is one-dimensional on segments, X is therefore not homogeneous, X is minimal under ϕ, ϕ is not almost periodic, ϕ is regularly almost periodic at some points of X but not at all points, ϕ is not recurrent and no orbits are unilaterally asymptotic.

There is also an example (F. B. Jones 1949) [**14**, p. 139 f.] of a minimal set under a discrete flow whose phase space is a plane one-dimensional continuum which is locally connected at some points and not locally connected at other points.

We turn now to another class of examples which we shall call *shifting flows*. Consider a nonvacuous compact metric space E and form its cartesian power $X=E^{\mathcal{I}}$ with exponent $\mathcal{I}$ where $\mathcal{I}$ is the set of all integers. An element of X is simply a function on $\mathcal{I}$ to E. Provide X with its product topology, or equivalently, its point-open topology. Define a homeomorphism σ of X onto X as follows: $(x_n | n\in\mathcal{I})\sigma = (x_{n+1} | n\in\mathcal{I})$ where $x=(x_n | n\in\mathcal{I})\in X$. The map σ is cutomarily called the *shift* transformation. A point x of X may be denoted by a so-called *symbolic trajectory*.

$$\begin{matrix} & & \downarrow & \\ \cdots x_{-1} & & x_0 x_1 & \cdots \end{matrix}$$

with *index* (the arrow) which denotes the value of the point x at 0. The shift transformation changes the indexed symbolic trajectory simply by shifting the index to the next symbol on the right. Thus

$$\begin{matrix} & & x & x\sigma & \\ & & \downarrow & \downarrow & \\ \cdots x_{-2} & x_{-1} & x_0 & x_1 & x_2 \cdots \end{matrix}$$

The notational device of a symbolic trajectory helps very considerably in the study of shifting flows.

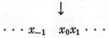

A *symbolic flow* is defined to be a shifting flow $(E^{\mathcal{I}}, \sigma)$ where E is a finite set with more than one element. The phase space $E^{\mathcal{I}}$ of a symbolic flow $(E^{\mathcal{I}}, \sigma)$ is the Cantor discontinuum. *Symbolic dynamics* is defined to be the study of symbolic flows; it is therefore largely combinatorial in nature. A large number of papers have been concerned with symbolic dynamics [**14**, p. 113]. The symbolic flows are descriptive of geodesic flows over certain surfaces of constant negative curvature. Marston Morse (1921) [**20**] defined a symbolic trajectory as follows, where $E=\{a, b\}$ with $a\neq b$. The elements a and b are called *duals* of each other. Consider the 1-block a; dualize to form the 1-block b and suffix it; we now have a 2-block $a\ b$. Consider the 2-block $a\ b$; dualize to form the 2-block $b\ a$ and suffix it; we now have a 4-block $a\ b\ b\ a$. The next step gives the 8-block $a\ b\ b\ a\ b\ a\ a\ b$. The process is continued to form a ray. This ray is then reflected to the left to form a trajectory which may be indexed at any symbol. This indexed trajectory is an example of an almost periodic point under σ which is not periodic. Consequently, Morse showed for the first time that there exist everywhere nonlocally connected compact minimal sets under certain geodesic flows.

A characteristic property of symbolic flows is *expansive*. A discrete flow (X, ϕ) is said to be *expansive* provided there exists a positive real number ϵ such that if x and y are different points of the phase space X, then $\rho(x\phi^n, y\phi^n)>\epsilon$ for some integer n. It may be seen that every symbolic flow is expansive and, conversely, if a discrete flow on a zero-dimensional compact metric space is expansive, then it may be imbedded in a symbolic flow. Observe that expansive is a strong negation of isometric, or equivalently, almost periodic. Consequently, the expansive minimal sets appear to be at the other end of the spectrum from almost periodic minimal sets. It has been proved (Schwartzmann 1952; Utz 1950 related) [**23**; **26**; **14**, p. 87] that every expansive discrete flow possesses a pair of points which is positively asymptotic and a pair of points which is negatively asymptotic. This means that $x\neq y$ and $\rho(x\phi^n, y\phi^n)\to 0$ as $n\to+\infty$ or $n\to-\infty$, respectively. The asymptotic points of a minimal set are at least partially characteristic of the minimal set.

In the shifting flow $(E^{\mathfrak{g}}, \sigma)$ we may take E to be the closed unit interval J and obtain the Hilbert cube $J^{\mathfrak{g}}$ as phase space. We may also take E to be the circle group K and obtain the infinite toral group $K^{\mathfrak{g}}$ as phase space; in this case the shift transformation is a group automorphism. An example (R. F. Williams 1955) [**28**] of an expansive discrete flow on a continuum may be imbedded in the flow $(K^{\mathfrak{g}}, \sigma)$. It is described as the set of all bisequences $(z_n/n\in\mathfrak{g})$ such that $z_n=z_{n+1}^2$ for all $n\in\mathfrak{g}$; it is therefore the 2-solenoidal group and the map is a group automorphism. The shifting flows in the Hilbert cube, the infinite torus, and other spaces, appear to be worthy of further study. So far as I am aware, shifting flows have been systematically investigated only in the case of symbolic flows.

I wish now to describe briefly some recent unpublished work [**3**, related] tending toward the classification of locally almost periodic minimal sets. Given a discrete flow (X, ϕ). The map ϕ is said to be *locally almost periodic* provided that if $x \in X$ and if U is a neighborhood of x, then there exists a neighborhood V of x and a relatively dense subset A of $\mathfrak{g}$ such that $V\phi^n \subset U$ for all $n \in A$. Two points x and y of X are said to be *proximal* provided that if $\epsilon > 0$, then $\rho(x\phi^n, y\phi^n) < \epsilon$ for some $n \in \mathfrak{g}$. Assume now that X is compact metric and ϕ is locally almost periodic. The relation proximal is a closed equivalence relation in X. Consider now the star-closed partition space X^* determined by the relation proximal. The discrete flow induced on X^* is almost periodic. Suppose also that X is minimal. Then X^* is minimal. Hence X^* is an almost periodic minimal set and thus a monothetic group, called the *structure group* of the original locally almost periodic minimal set. To illustrate, the structure group of Floyd's example is the triadic group, and the structure group of Jones' example in the circle group.

Just a few words now about continuous flows. In general, the definitions and theorems mentioned before apply to continuous flows as well as discrete flows. There are some exceptions, however. A minimal set under a continuous flow is necessarily connected; not so for a discrete flow. Another exception appears in the theorem (A. A. Markov 1931) [**19**; **14**, p. 14] that a finite-dimensional minimal set under a continuous flow is a Cantor-manifold and hence has the same dimension at each point. This is not true for discrete flows: witness the example of Floyd.

The examples of continuous flows which have been most extensively studied are the geodesic flows over surfaces of constant negative curvature. Geodesic flows are defined for Riemannian manifolds of arbitrary dimension but let us think of the 2-dimensional case. Let S be a surface of class C^2 provided with a Riemannian metric of class C^2 and which is complete in the sense that each geodesic arc in S can be extended to infinite length in both directions. The phase space X consists of the unit contravariant vectors on S which we may call the unitangents (unitangent = unit tangent). Let $x \in X$ and let $t \in \mathfrak{R}$. Consider the geodesic g in S which has x as a tangent. If t is positive, measure t units of arc length along g in the same direction as x from the base point of x and take the unitangent to g at the new point to be the image of x under the transition π^t. Likewise for negative t. There results a continuous flow which is called the *geodesic flow over S*. Now X is a 3-dimensional manifold which is the bundle space of a fibre bundle with the circle as fiber and the surface S as base space. The geodesic flows are dynamical systems which are determined by the Euler equations.

We state one theorem on geodesic flows which is now classical. If S is a complete 2-dimensional analytic Riemannian manifold of constant negative curvature and of finite area, then the geodesic flow over S is regionally mixing and its periodic points are dense in the space of unitangents on S. This theorem is proved again in the Colloquium volume *Topological dynamics* [**14**, p. 131]. A continuous flow $(X, \mathcal{R}, \pi)$ is said to be *regionally mixing* provided that if U and V are nonvacuous open subsets of X, then there exists a positive real number s such that $Ut \cap V \neq \emptyset$ for $t \in \mathcal{R}$ with $|t| \geqq s$.

Closely related to the geodesic flows are the horocycle flows over surfaces of constant negative curvature. They have the same phase spaces as geodesic flows but different transitions and motions. It has been shown (Hedlund 1936) [**16**] that for each integer $p > 1$ there exists a closed orientable surface S of constant negative curvature and of genus p such that the space X of unitangents on S is a regionally mixing minimal set under the horocycle flow over S. Here the phase spaces X are compact 3-dimensional manifolds, are not almost periodic under the horocycle flows, and, even more, are not recurrent (since regionally mixing). A continuous flow $(X, \mathcal{R}, \pi)$ is said to be *recurrent* provided that if $\epsilon > 0$, then there exists an extensive subset A of $\mathcal{R}$ such that $xt \in N_\epsilon(x)$ for all $x \in X$ and all $t \in A$.

As remarked previously, every discrete flow (X, ϕ) gives rise to a continuous flow as follows: take the cartesian product of X and the closed unit interval, and identify bases according to the map ϕ. The capped n-adic minimal sets produce the n-solenoidal minimal sets. All of these minimal sets are regularly almost periodic, and their structure groups are the n-adic groups and the n-solenoidal groups.

Consider the space X of all continuous functions f on the line group R to $\mathcal{R}$ and provide X with its compact-open topology. Define a continuous flow in X by translation of the function f: $(f(r) | r \in \mathcal{R}) \rightarrow (f(r+t) | r \in \mathcal{R})$ defines π^t for each $t \in \mathcal{R}$. It is known (Bebutov 1940) [**4**] that every continuous flow in a compact metric space with at most one fixed point is imbeddable in the above flow. The minimal sets under continuous flows therefore all appear in the Bebutov example. A perspicuous proof of his theorem is much to be desired. From a topological point of view, the Bebutov example has a favorable property, namely, the orbit-closure of a point f is compact if and only if f is bounded and uniformly continuous.

The consideration of function spaces leads us to the almost periodic functions of Harold Bohr. Consider the space X of all bounded uniformly continuous functions on $\mathcal{R}$ to $\mathcal{R}$ and provide X with its uniform topology. Define a continuous flow in X by translation of the functions. An element of X is an almost periodic function in the sense of Bohr if and only if it is an almost periodic point under the flow. This statement shows the consistency of the terminology almost periodic. A certain fraction of the theory of almost periodic functions can be presented under the theory of continuous flows or, more generally, transformation groups, [**14**, pp. 40–48]. It seems that future developments will encompass more. Recent work of Tornehave (1954) [**25**] is suggestive in this connection.

In the foregoing discussion of flows there are several features which suggest a broader and more general development in the context of transformation groups. Some of these features are: (1) analogous theories for discrete flows and continuous flows (the famous dictum of E. H. Moore is brought to mind); (2) the study of flows themselves leads to topological groups; (3) the study of flows leads to almost periodic functions and the theory of almost periodic functions includes functions on groups.

The Colloquium volume *Topological dynamics* [**14**] is written from this more general point of view. Let us mention briefly two representative notions of recursion for transformation groups. Let (X, T) be a transformation group. A subset A of the phase group T is said to be *syndetic* provided that $T=AK$ for some compact subset K of T. The notion syndetic replaces the notion relatively dense for sets in $\mathcal{I}$ and $\mathcal{R}$. A point x of the phase space X is said to be *almost periodic* provided that if U is a neighborhood of x, then $xA=\{xt \mid t\in A\}\subset U$ for some syndetic subset A of T. A semigroup P in T is said to be *replete* provided that P contains some bilateral translate of each compact subset of T. A subset A of T is said to be *extensive* provided that A intersects every replete semigroup in T. A point x of X is said to be *recurrent* provided that if U is a neighborhood of x, then $xA\subset U$ for some extensive subset A of T. When T is $\mathcal{I}$ or $\mathcal{R}$, then these notions of recursion reduce to the customary ones for flows [**14**].

For the reamining moments permit me to pose a few more questions which are presently unanswered so far as I know. Questions stated for flows on metric spaces are usually meaningful also for transformation groups on more general spaces.

(1) What compact metric spaces can be minimal sets under a discrete flow? Under a continuous flow? The universal curve of Sierpinski? The universal curve of Menger? A lens space? What polyhedra? Can they be nonorientable? About all that is known for polyhedra is that, in the continuous flow case, the Euler characteristic has to vanish. This follows from the fact that a minimal set with more than one point cannot have the fixed point property [**1**, p. 532]. The only closed surfaces with vanishing Euler characteristic are the torus

and the Klein bottle. The torus is a minimal set under both discrete flows and continuous flows. Can the Klein bottle be a minimal set? What can be said about the homology and homotopy groups of a minimal set?

(2) More generally, what subsets of a given phase space can be minimal subsets? Does there exist a discrete flow in euclidean 3-space such that some orbit-closure is the necklace of Antoine? If so, can it be minimal? In an n-dimensional manifold it is known that a minimal subset is either the whole manifold or of dimension less than n [**14**, p. 14]. If the dimension is equal to $n-1$, then is it necessarily an $(n-1)$-torus? (Question of R. W. Bass.) Bass has pointed out that a theorem of Kodaira and Abe (1940) [**18**] shows the answer to be affirmative for an almost periodic minimal set under a continuous flow even when merely the phase space is imbeddable in euclidean n-space.

(3) Let X be a compact uniform space which is minimal under a homeomorphism ϕ. In general, X need not be metrizable since every compact connected separable abelian group is monothetic (for example, K^J where K is the circle group and J is the unit interval). What other conditions will guarantee that X be metrizable? One instance is that ϕ be expansive (Bryant 1955) [**7**; **9**]. Another is that ϕ be regularly almost periodic.

(4) Distal means without distinct proximal points. Is every distal minimal set necessarily almost periodic? It is known that distal and locally almost periodic are equivalent to almost periodic, even without minimality [**13**].

(5) The study of geodesic flows and horocycle flows now requires rather much geometry and analysis. Is it possible, in some sense, to axiomatize these flows so that they are more accessible to immediate study?

(6) In many examples of geodesic flows, symbolic flows, and horocycle flows the following properties occur together: regionally mixing and dense periodic or almost periodic points. Does regionally mixing plus some auxiliary hypothesis imply that the almost periodic points are dense?

(7) What spaces can carry an expansive homeomorphism (Question of Utz and others)? The n-cells? May such spaces be locally connected?

(8) The roles of connected, locally connected, locally euclidean, and other kinds of connectivity, in question of existence and implication?

(9) When von Neumann defined his almost periodic functions on a

group, he generalized Bochner's characterization of the Bohr almost periodic functions. This led to the notion of left and right almost periodic functions which Maak subsequently proved equivalent. (Incidentally, the theorems of topological dynamics can be quoted to establish this.) When the original Bohr definition is generalized directly to functions on topological groups, a similar dichotomy occurs but the answer as to their equivalence in this case is not clear. If the two kinds are actually equivalent, then this new definition leads again to von Neumann's functions. (See [**14**, p. 40 f.] for references and full statements.)

(10) There are a number of theorems in topological dynamics whose hypothesis and conclusion are meaningful for nonmetric spaces, say compact or locally compact uniform spaces, and yet the only known proof makes use of a category argument valid for metric spaces or at least first-countable spaces [**14**]. The question is to find a replacement for the category arguments, either in particular cases or in general, which permit the removal of countability assumptions such as metric or first-countable. Recent work of Robert Ellis (1957) [**10**] is very interesting and suggestive in this connection.

(11) From an abstract point of view, there is an equivalence between transformation groups and spaces of functions on groups to spaces. Briefly, the space X of functions on a group T to a space Y determines a transformation group (X, T) by translation on the functions; and a transformation group (X, T, π) determines the space $\{\pi_x | x \in X\}$ of motions in the space Z of all functions on T to X. These two constructions are not symmetric because the first produces a relatively complicated space X from a relatively simple space Y, and the second produces a more complicated space Z from an already complicated space X. To avoid the difficulty, a theorem is needed for transformation groups which extends the Bebutov theorem on a universal flow for continuous flows.

More questions are stated in [**21**].

References

1. Paul Alexandroff and Heinz Hopf, *Topologie* I, Die Grundlehren der mathematischen Wissenschaften, vol. 45, Berlin, 1935; Ann Arbor, 1945.

2. Hirotada Anzai and S. Kakutani, *Bohr compactifications of a locally compact abelian group* II, Proceedings of the Imperial Academy, Tokyo, vol. 19 (1943) pp. 533–539.

3. Joseph Auslander, *Mean-L-stable systems*, Dissertation, University of Pennsylvania, 1957.

4. M. Bebutov, *Sur les systèmes dynamiques dans l'espace des fonctions continues*, Comptes Rendus, Doklady Akad. Nauk SSSR, vol. 27 (1940) pp. 904–906.

5. G. D. Birkhoff, *Dynamical systems*, Amer. Math. Soc. Colloquium Publications, vol. 9, New York, 1927.

6. ———, *Collected mathematical papers*, 3 vols., American Mathematical Society, New York, 1950.

7. B. F. Bryant, *Unstable self-homeomorphisms of a compact space*, Dissertation, Vanderbilt University, 1954.

8. ———, *A note on unstable homeomorphisms*, Amer. Math. Monthly vol. 61 (1954) p. 509 (abstract).

9. ———, *A note on uniform spaces*, Amer. Math. Monthly, vol. 62 (1955) p. 529 (abstract).

10. Robert Ellis, *Locally compact transformation groups*, Duke Math. J. vol. 24 (1957) pp. 119–126.

11. E. E. Floyd, *A nonhomogeneous minimal set*, Bull. Amer. Math. Soc. vol. 55 (1949) pp. 957–960.

12. W. H. Gottschalk, *Almost periodic points with respect to transformation semi-groups*, Ann. of Math. vol. 47 (1946) pp. 762–766.

13. ———, *Characterizations of almost periodic transformation groups*, Proc. Amer. Math. Soc. vol. 7 (1956) pp. 709–712.

14. W. H. Gottschalk and G. A. Hedlund, *Topological dynamics*, Amer. Math. Soc. Colloquium Publications, vol. 36, Providence, 1955.

15. P. R. Halmos and H. Samelson, *On monothetic groups*, Proc. Nat. Acad. Sci. U.S.A., vol. 28 (1942) pp. 254–258.

16. G. A. Hedlund, *Fuchsian groups and transitive horocycles*, Duke Math. J. vol. 2 (1936) pp. 530–542.

17. E. R. van Kampen, *The topological transformations of a simple closed curve into itself*, Amer. J. Math. vol. 57 (1935) pp. 142–152.

18. Kunihiko Kodaira and Makoto Abe, *Uber zusammenhängende kompakte abelsche Gruppen*, Proceedings of the Imperial Academy, Tokyo, vol. 16 (1940) pp. 167–172.

19. A. A. Markoff, *Sur une propriété générale des ensembles minimaux de M. Birkhoff*, C. R. Acad. Sci. Paris vol. 193 (1931) pp. 823–825.

20. Marston Morse, *Recurrent geodesics on a surface of negative curvature*, Trans. Amer. Math. Soc. vol. 22 (1921) pp. 84–100.

21. V. V. Nemyckiĭ, *Topological problems of the theory of dynamical systems*, Uspehi Mat. Nauk (N.S.) vol. 4 (1949) pp. 91–152 (Russian); Amer. Math. Soc. Translation, number 103.

22. V. V. Nemyckiĭ and V. V. Stepanov, *Qualitative theory of differential equations*, Moscow-Leningrad, 1947; 2d ed., 1949, Russian; English translation of 2d ed., Princeton University Press, to appear in 1959.

23. Sol Schwartzman, *On transformation groups*, Dissertation, Yale University, 1952.

24. P. A. Smith, *Periodic and nearly periodic transformations*, Lectures in Topology, Ann Arbor, 1941, pp. 159–190.

25. Hans Tornehave, *On almost periodic movements*, Det Kongelige Danske Videnskabernes Selskab, Matematisk-fysiske Meddelelser, vol. 28, no. 13, 1954.

26. W. R. Utz, *Unstable homeomorphisms*, Proc. Amer. Math. Soc. vol. 1 (1950) pp. 769–774.

27. G. T. Whyburn, *Analytic topology*, Amer. Math. Soc. Colloquium Publications, vol. 28, New York, 1942.

28. R. F. Williams, *A note on unstable homeomorphisms*, Proc. Amer. Math. Soc. vol. 6 (1955) pp. 308–309.

UNIVERSITY OF PENNSYLVANIA

BIBLIOGRAPHY

ANZAI, HIROTADA

1. (With S. Kakutani) *Bohr compactifications of a locally compact abelian group* I, Proceedings of the Imperial Academy, Tokyo, vol. 19 (1943), pp. 476–480.
2. (With S. Kakutani) *Bohr compactifications of a locally compact abelian group* II, Proceedings of the Imperial Academy, Tokyo, vol. 19 (1943), pp. 533–539.

ARENS, RICHARD

1. *A topology for spaces of transformations*, Annals of Mathematics, vol. 47 (1946), pp. 480–495.
2. *Topologies for homeomorphism groups*, American Journal of Mathematics, vol. 68 (1946), pp. 593–610.

AYRES, W. L.

1. *On transformations having periodic properties*, Fundamenta Mathematicae, vol. 33 (1939), pp. 95–103.

BAUM, J. D.

1. *An equicontinuity condition for transformation groups*, Proceedings of the American Mathematical Society, vol. 4 (1953), pp. 656–662.

BEBUTOFF, M.

1. *Sur les systèmes dynamiques dans l'espace des fonctions continues*, Comptes Rendus (Doklady) de l'Académie des Sciences de l'URSS, vol. 27 (1940), pp. 904–906.

BELTRAMI, E.

1. *Teoriá fondamentale degli spazii di curvatura constante*, Annali di Matematica, ser. 2, vol. 2, pp. 232–255.

BESICOVITCH, A. S.

1. *A problem on topological transformations of the plane*, Fundamenta Mathematicae, vol. 28 (1937), pp. 61–65.
2. *A problem on topological transformations of the plane*. II, Proceedings of the Cambridge Philosophical Society, vol. 47 (1951), pp. 38–45.

BIANCHI, L.

1. *Vorlesungen über Differentialgeometrie*, Leipzig, 1899.

BIRKHOFF, G. D.

1. *Collected Mathematical Papers*, vols. 1, 2, 3, New York, 1950.
2. *Dynamical Systems*, American Mathematical Society Colloquium Publications, vol. 9, 1927.

BOCHNER, S.

1. *Beiträge zur Theorie der fastperiodischen Funktionen* I, Mathematische Annalen, vol. 96 (1926), pp. 119–147.

BOHR, HARALD

1. *Collected Mathematical Works*, vols. 1, 2, 3, Copenhagen, 1952.

BOURBAKI, N.

Eléments de mathématique, Première Partie, *Les structures fondamentales de l'analyse.*

1. Livre I, *Théorie des ensembles* (*Fasicule de resultats*), Actualités scientifiques et industrielle 846, Paris, 1939.
2. Livre III, *Topologie générale*, Chapitre I, *Structures topologiques*, & Chapitre II, *Structures uniformes*, Actualités scientifiques et industrielles 858, Paris, 1940.
3. Livre III, *Topologie générale*, Chapitre III, *Groupes topologiques* (*Théorie élémentaire*), Actualités scientifiques et industrielles 916, Paris, 1942.
4. Livre III, *Topologie générale*, Chapitre X, *Espaces fonctionnels*, Actualités scientifiques et industrielles 1084, Paris, 1949.

CAMERON, R. H.

1. *Almost periodic transformations*, Transactions of the American Mathematical Society, vol. 36 (1934), pp. 276–291.

CARATHÉODORY, C.

1. *Über den Wiederkehrsatz von Poincaré*, Sitzungsberichte der Preussischen Akademie der Wissenschaften, 1919, pp. 580–584.

ELLIS, ROBERT

1. *Continuity and homeomorphism groups*, Proceedings of the American Mathematical Society, vol. 4 (1953), pp. 969–973.

ENGEL, F.

1. See Lie [1].

ERDÖS, P.

1. (With A. H. Stone) *Some remarks on almost periodic transformations*, Bulletin of the American Mathematical Society, vol. 51 (1945), pp. 126–130.

FLOYD, E. E.

1. A *nonhomogeneous minimal set*, Bulletin of the American Mathematical Society, vol. 55 (1949), pp. 957–960.

FORD, L. R.

1. *Automorphic Functions*, New York, 1929.

FORT, M. K., Jr.

1. *A note on equicontinuity*, Bulletin of the American Mathematical Society, vol. 55 (1949), pp. 1098–1100.

FRANKLIN, PHILIP

1. *Almost periodic recurrent motions*, Mathematische Zeitschrift, vol. 30 (1929), pp. 325–331.

FRICKE, R.

1. (With F. Klein) *Automorphe Functionen*, vol. I, Leipzig, 1897.

GARCIA, MARIANO

1. (With G. A. Hedlund) *The structure of minimal sets*, Bulletin of the American Mathematical Society, vol. 54 (1948), pp. 954–964.

GOTTSCHALK, W. H.

1. *An investigation of continuous mappings with almost periodic properties*, Dissertation, University of Virginia, 1944.
2. *Powers of homeomorphisms with almost periodic properties*, Bulletin of the American Mathematical Society, vol. 50 (1944), pp. 222–227.
3. *Orbit-closure decompositions and almost periodic properties*, Bulletin of the American Mathematical Society, vol. 50 (1944), pp. 915–919.
4. *A note on pointwise nonwandering transformations*, Bulletin of the American Mathematical Society, vol. 52 (1946), pp. 488–489.
5. (With G. A. Hedlund) *Recursive properties of transformation groups*, Bulletin of the American Mathematical Society, vol. 52 (1946), pp. 637–641.
6. *Almost periodic points with respect to transformation semi-groups*, Annals of Mathematics, vol. 47 (1946), pp. 762–766.
7. *Almost periodicity, equi-continuity and total boundedness*, Bulletin of the American Mathematical Society, vol. 52 (1946), pp. 633–636.
8. *Recursive properties of transformation groups* II, Bulletin of the American Mathematical Society, vol. 54 (1948), pp. 381–383.
9. *Transitivity and equicontinuity*, Bulletin of the American Mathematical Society, vol. 54 (1948), pp. 982–984.
10. (With G. A. Hedlund) *The dynamics of transformation groups*, Transactions of the American Mathematical Society, vol. 65 (1949), pp. 348–359.
11. *The extremum law*, Proceedings of the American Mathematical Society, vol. 3 (1952), p. 631.

HALL, D. W.

1. (With J. L. Kelley) *Periodic types of transformations*, Duke Mathematical Journal, vol. 8 (1941), pp. 625–630.

HALMOS, P. R.

1. (With H. Samelson) *On monothetic groups*, Proceedings of the National Academy of Sciences of the United States, vol. 28 (1942), pp. 254–258.
2. *Measure Theory*, New York, 1950.
3. (With H. E. Vaughan) *The marriage problem*, American Journal of Mathematics, vol. 72 (1950), pp. 214–215.

HARTMAN, PHILIP

1. (With Aurel Wintner) *Integrability in the large and dynamical stability*, American Journal of Mathematics, vol. 65 (1943), pp. 273–278.

HAUSDORFF, FELIX

1. *Grundzüge der Mengenlehre*, 1st ed., reprinted, Chelsea, New York, 1949.

HEDLUND, G. A.

1. See Morse [3].
2. *The dynamics of geodesic flows*, Bulletin of the American Mathematical Society, vol. 45 (1939), pp. 241–260.
3. See Morse [4].
4. *Sturmian minimal sets*, American Journal of Mathematics, vol. 66 (1944), pp. 605–620.
5. See Gottschalk [5].
6. See Garcia [1].
7. See Gottschalk [10].

HILMY, HEINRICH

1. *Sur une propriété des ensembles minima*, Comptes Rendus (Doklady) de l'Académie des Sciences de l'URSS, vol. 14 (1937), pp. 261–262.
2. *Sur la théorie des ensembles quasi-minimaux*, Comptes Rendus (Doklady) de l'Académie des Sciences de l'URSS, vol. 15 (1937), pp. 113–116.
3. *Sur les mouvements des systèmes dynamiques qui admettent "l'incompressibilite" des domaines*, American Journal of Mathematics, vol. 59 (1937), pp. 803–808.
4. *Sur les théorèmes de récurrence dans la dynamique générale*, American Journal of Mathematics, vol. 61 (1939), pp. 149–160.

HOPF, E.

1. *Zwei Sätze über den wahrscheinlichen Verlauf der Bewegungen dynamischer Systeme*, Mathematische Annalen, vol. 103 (1930), pp. 710–719.
2. *Ergodentheorie*, Berlin, 1937.
3. *Statistik der geodätischen Linien in Mannigfaltigkeiten negativer Krümmung*, Berichte über die Verhandlungen der sächsischen Akademie der Wissenschaften zu Leipzig, Mathematisch-Physischen Klasse, vol. 91 (1939), pp. 261–304.

HOPF, H.

1. *Zum Clifford-Kleinschen Raumproblem*, Mathematische Annalen, vol. 95 (1926), pp. 313–339.
2. (With W. Rinow) *Ueber den Begriff der vollständigen differentialgeometrische Fläche*, Commentarii Mathematici Helvetici, vol. 3 (1931), pp. 209–225.

HUREWICZ, WITOLD

1. (With Henry Wallman) *Dimension Theory*, Princeton Mathematical Series No. 4, 1941.

KAKUTANI, S.

1. See Anzai [1].
2. See Anzai [2].

KELLEY, J. L.

1. See Hall [1].

von Kerékjártó, B.

1. *Sur les similitudes de l'espace*, Comptes Rendus de l'Académie des Sciences Paris, vol. 198 (1934), p. 1345.

Klein, F.

1. See Fricke [1].

Koebe, P.

1. *Riemannsche Mannigfaltigkeiten und nicht euklidische Raumformen*, Sitzungsberichte der Preussischen Akademie der Wissenschaften, (1927), pp. 164–196; (1928), pp. 345–442; (1929), pp. 414–457; (1930), pp. 304–364, 504–541; (1931), pp. 506–534.

Kuratowski, Casimir

1. *Topologie* I, Warsaw, 1933.

Kurosch, A.

1. *Theory of Groups*, Moscow, 1944, (Russian).

Lie, Sophus

1. (With F. Engel) *Theorie der Transformationgruppen*, Leipzig, 1888.

Löbell, F.

1. *Die überall regulären unbegrenzten Flächen fester Krümmung*, Dissertation, Tübungen, 1927.
2. *Beispiele geschlossener dreidimensionaler Clifford-Kleinscher Räume negativer Krümmung*, Berichte über die Verhandlungen der sächischen Akademie der Wissenschaften zu Leipzig, Mathematisch-Physischen Klasse, vol. 83 (1931), pp. 167–174.

Maak, Wilhelm

1. *Eine neue Definition der fastperiodischen Funktionen*, Abhandlungen aus dem Mathematischen Seminar der Hansischen Universität, vol. 11 (1936), pp. 240–244.
2. *Abstrakte fastperiodische Funktionen*, Abhandlungen aus dem Mathematischen Seminar der Hansischen Universität, vol. 11 (1936), pp. 367–380.

Maier, A. G.

1. *On central trajectories and a problem of Birkhoff*, Matematičeskiĭ Sbornik (N. S.), vol. 26 (68), (1950), pp. 266–290.

Markoff, A. A.

1. *Sur une propriété générale des ensembles minimaux de M. Birkhoff*, Comptes Rendus de l'Académie des Sciences, Paris, vol. 193 (1931), pp. 823–825.
2. *Stabilität im Liapounoffschen Sinne und Fastperiodizität*, Mathematische Zeitschrift, vol. 36 (1933), pp. 708–738.

Montgomery, Deane

1. *Almost periodic transformation groups*, Transactions of the American Mathematical Society, vol. 42 (1937), pp. 322–332.
2. *Pointwise periodic homeomorphisms*, American Journal of Mathematics, vol. 59 (1937), pp. 118–120.
3. *Measure preserving homeomorphisms at fixed points*, Bulletin of the American Mathematical Society, vol. 51 (1945), pp. 949–953.

Morse, Marston

1. *A one-to-one representation of geodesics on a surface of negative curvature*, American Journal of Mathematics, vol. 43 (1921), pp. 33–51.
2. *Recurrent geodesics on a surface of negative curvature*, Transactions of the American Mathematical Society, vol. 22 (1921), pp. 84–100.
3. (With G. A. Hedlund) *Symbolic dynamics*, American Journal of Mathematics, vol. 60 (1938), pp. 815–866.
4. (With G. A. Hedlund) *Symbolic dynamics* II. *Sturmian trajectories*, American Journal of Mathematics, vol. 62 (1940), pp. 1–42.

Niemytzki, V. V.

1. (With V. V. Stepanoff) *Qualitative Theory of Differential Equations*, Moscow-Leningrad, 1st ed. 1947, 2nd ed. 1949, (Russian).

2. *Topological problems of the theory of dynamical systems*, Uspehi Matematiceskih Nauk (N. S.), vol. 4 (1949), pp. 91–153, (Russian); American Mathematical Society, Translation Number 103.

von Neumann, J.

1. *Almost periodic functions in a group*. I, Transactions of the American Mathematical Society, vol. 36 (1934), pp. 445–492.

Oxtoby, J. C.

1. *Note on transitive transformations*, Proceedings of the National Academy of Sciences of the United States, vol. 23 (1937), pp. 443–446.
2. (With S. M. Ulam) *Measure preserving homeomorphisms and metric transitivity*, Annals of Mathematics, vol. 42 (1941), pp. 874–920.
3. *Ergodic sets*, Bulletin of the American Mathematical Society, vol. 58 (1952), pp. 116–136.

Rinow, W.

1. See H. Hopf [2].

Salenius, Tauno

1. *Über dreidimensionale geschlossene Räume konstanter negativer Krümmung*, Den 11te Skandinaviske Matematikerkongress, Trondheim, 1949, pp. 107–112.

Samelson, Hans

1. See Halmos [1].

Schwartzman, S.

1. *On transformation groups*, Dissertation, Yale University, 1952.

Seidel, W.

1. (With J. L. Walsh) *On approximation by euclidean and non-euclidean translations of an analytic function*, Bulletin of the American Mathematical Society, vol. 47 (1941), pp. 916–920.

Smith, P. A.

1. *Periodic and nearly periodic transformations*, Lectures in Topology, Ann Arbor, 1941, pp. 159–190.

Stepanoff, V. V.

1. (With A. Tychonoff) *Über die Räume der fastperiodischen Funktionen*, Recueil Mathématique, Nouvelle Série, vol. 41 (1934), pp. 166–178.
2. See Niemytzki [2].

Stone, A. H.

1. See Erdös [1].

Tychonoff, A.

1. See Stepanoff [1].

Ulam, S. M.

1. See Oxtoby [2].

Utz, W. R.

1. *Unstable homeomorphisms*, Proceedings of the American Mathematical Society, vol. 1 (1950), pp. 769–774.

Vaughan, H. E.

1. See Halmos [3].

Wallman, H.

1. See Hurewicz [1].

Walsh, J. L.

1. See W. Seidel [1].

Weil, André

1. *L'intégration dans les groupes topologiques et ses applications*, Paris, 1940.

Whyburn, G. T.

1. *Analytic Topology*, American Mathematical Society Colloquium Publications, vol. 28, New York, 1942.

WILLIAMS, C. W.

1. *Recurrence and incompressibility*, Proceedings of the American Mathematical Society, vol. 2 (1951), pp. 798–806.

WINTNER, A.

1. See Hartman [1].

ZIPPIN, LEO

1. *Transformation groups*, Lectures in Topology, Ann Arbor, 1941, pp. 191–221.

Mathematics in U. S. S. R. for thirty years, 1917–1947, Moscow-Leningrad, 1948, pp. 508–517.

A limited number of copies of the *Bibliography for Dynamical Topology*, Fifth Edition, September 1972, compiled by Walter H. Gottschalk, Wesleyan University, is available (at cost) from the Department of Mathematics, Wesleyan University, Middletown, Connecticut 06457. This Bibliography contains 178 pages and lists 2475 numbered items.

INDEX

DEFGHIJ–CM–898765432